Working Skills in
Geometric Dimensioning and Tolerancing

Working Skills in
Geometric Dimensioning and Tolerancing

Mike Fitzpatrick

Delmar Publishers Inc.

NOTICE TO THE READER

Publisher does not warrant or guarantee any of the products described herein or perform any independent analysis in connection with any of the product information contained herein. Publisher does not assume, and expressly disclaims, any obligation to obtain and include information other than that provided to it by the manufacturer.

The reader is expressly warned to consider and adopt all safety precautions that might be indicated by the activities described herein and to avoid all potential hazards. By following the instructions contained herein, the reader willingly assumes all risks in connection with such instructions.

The publisher makes no representations or warranties of any kind, including but not limited to, the warranties of fitness for particular purpose or merchantability, nor are any such representations implied with respect to the material set forth herein, and the publisher takes no responsibility with respect to such material. The publisher shall not be liable for any special, consequential or exemplary damages resulting, in whole or in part, from the readers' use of, or reliance upon, this material.

Cover design courtesy of: Juan Vargas/Vargas Williams Designs

Delmar Staff
    Executive Editor: Michael McDermott
    Associate Editor: Kevin Johnson
    Project Editor: Carol Micheli
    Production Supervisor: Wendy Troeger
    Art Supervisor: Judi Orozco
    Design Supervisor: Susan C. Mathews

For information, address Delmar Publishers Inc.
3 Columbia Circle, Box 15-015
Albany, New York 12212

Copyright © 1993
by Delmar Publishers Inc.

All rights reserved. No part of this work covered by the copyright may be reproduced or used in any form or by any means—graphic, electronic, or mechanical, including photocopying, recording, taping, or information storage and retrieval systems—without written permission of the publisher.

printed in the United States of America
published simultaneously in Canada
by Nelson Canada,
a division of The Thomson Corporation

1 2 3 4 5 6 7 8 9 10 XXX 99 98 97 96 95 94 93

Library of Congress Cataloging-in-Publication Data
    Fitzpatrick, Michael, 1945–
      Working skills in geometric dimensioning and tolerancing / Mike Fitzpatrick.
        p. cm.
      Includes index.
      ISBN 0-8273-4900-9 (textbook)
      1. Engineering drawings—Dimensioning. 2. Tolerance (Engineering)
    I. Title
    T357.F68 1993
    620'.0045—dc20
                                                               92-18034
                                                                    CIP

Dedication

This book is a dedicated to the

Brown and Sharpe Corporation

— For the help they freely gave in making this book possible
— For their history of being at the "root of precision" in American manufacturing
— For their efforts to improve the future of manufacturing
— Especially for their annual Metrology Grant whereby they have shown confidence in vocational and technical education

With insight and committment, Brown and Sharpe is investing in the future of manufacturing by planting "seeds of quality" within education.
Thank you.

Contents

Preface   xv
Acknowledgements   xvii

UNIT I
FOUNDATION SKILLS

Unit Introduction   2

CHAPTER 1 What is the Geometric Dimensioning and Tolerancing System?   3

    A Tolerancing Control System   3
    Statistical Process Control and Geometrics   5
    Five Types of Feature Control   6

CHAPTER 2 What Advantages Are There to Geometrics?   8

    The Industrial Need for Geometrics   8
    A Controlled/Supervised System   8
    Advantages to the Geometric System   9
    Symbols Used in Geometric Dimensioning and Tolerancing   10
    Tolerance Versus Cost   10

UNIT II
THE WORKING GEOMETRIC CONCEPTS

Unit Introduction   14

CHAPTER 3 Datums—A Basis For Measuring and Positioning   15

   A Starting Point for Dimensioning, Measuring, and Positioning   15
   Datum Features Establish Datums   17
   Datum Callout Symbols and Order of Importance   19
   Datum Priorities per Design Function   20
   Datum Target Symbols   23
   Datum Frames   26
   Challenge Problem 3-1   30
   Answers to Challenge Problem 3-1   31
   Review of the Concept of Datums   31

CHAPTER 4 Interpreting Feature Control Frames   33

   Single Example   33
   Combined Example   34
   Multiple Example   34
   Information Found in Frames   35
   Scanning a Print   36
   Challenge Problem 4-1   37
   Answers to Challenge Problem 4-1   38

CHAPTER 5 Geometric Tolerance Zones—the Control Tool   39

   What Entities are Found in the Tolerance Zone   39
   Two- and Three-dimensional Surface Controls   40

Contents □ ix

CHAPTER 6 Controls of Form and Profile — 42

Controls of Form and Profile Seek to Confine Surface Within
  a Tolerance Zone — 42
Two- and Three-dimensional Controls — 43
All Controls of Surface Form are Floating Template
  Controls—Concept #4 — 43
Computers Test Form Mathematically — 44
Straightness—Concept #5 — 44
Straightness Review — 49
Flatness—Concept #6 — 51
Flatness Review — 56
Challenge Problem 6-1 — 57
Answers to Challenge Problem 6-1 — 57
Roundess—Concept #7 — 57
Roundness Review — 61
Cylindricity—Concept #8 — 62
Cylindricity Review — 63
Controls of Profile — 64
Profile of a Line—Concept #9 — 67
Profile of a Surface—Concept #10 — 71
Review of Controls of Profile — 72

CHAPTER 7 Controls of Orientation—Parallelism, Angularity, and Perpendicularity — 74

Parallelism—Concept #11 — 75
Perpendicularity—Concept #12 — 78
Angularity—Concept #13 — 82
Challenge Problem 7-1 — 86
Answers to Challenge Problem 7-1 — 86
Orientation Review — 86

CHAPTER 8 Material Modifiers — 88

The Material Conditions — 92
Maximum Material Condition—Concept #17 — 93
Least Material Condition—Concept #18 — 93
Challenge Problem 8-1 — 93

Answers to Challenge Problem 8-1   95
The Earned Tolerance Calculation Procedure—Concept #19   96
Challenge Problem 8-2   97
Answers to Challenge Problem 8-2   100
Regardless of Feature Size—Concept #20   100
Review of Material Condition Modifiers   102

CHAPTER 9 Characteristic Controls of Runout   103

Circular Runout—Concept #21   103
Total Circular Runout—Concept #22   103
Review of Runout   107

CHAPTER 10 Characteristics of Location   109

Concentricity—Concept #23   109
Review of Concentricity   116
Control of Position—Concept #24   116
Formula for Finding Earned Tolerance   121
Inspecting Geometric Position   122
Position/Concentricity Challenge Problem 10-1   124
Answers to Challenge Problem 10-1   127
Review of Position   128

CHAPTER 11 Geometric System Notes   129

Projected Tolerance Zone—Concept #28   129
Zero Position Tolerance at MMC—Concept #29   130
Composite Positional Tolerances   132
Conical and Bidirectional Tolerancing   134
Variation—Free State or Restrained   135

CHAPTER 12 A Systems Approach to Geometrics   137

Combining Controls   137
The 13 Characteristics   140

Unit II Self Challenge Problem 12-1	140
Answers to Challenge Problem 12-1	144

UNIT III
APPLICATION SKILLS IN MANUFACTURING

Unit Introduction	148

CHAPTER 13 Single Axis Feature Inspection and Rework — 149

Inspection of Position	149
Challenge Problem 13-1	151
Answers to Challenge Problem 13-1	154
Reworking Single Axis Features	156
Reworking Features Where No Bonus Tolerance Applies	157
Rule of Sides	157
Challenge Problem 13-2	160
Answers to Challenge Problem 13-2	162
Reworking Features Where Bonus Tolerances Apply	163
MMC Rework Formula	166
Challenge Problem 13-3	166
Answers to Challenge Problem 13-3	169
LMC Method to See if Rework is Possible	169
LMC Test Formula	170
Review of Inspection and Rework of Single Axis Features	170

CHAPTER 14 Inspecting and Reworking Two Axis Features — 172

Converting from Rectangular to Geometric Position	172
Computing a Balance Sheet for a Hole or Other Round Feature—Steps 1, 2, and 3	175
Pythagorean's Theorem	176
Challenge Problem 14-1	177
Answers to Challenge Problem 14-1	180

Determining Rework Amount	181
Challenge Problem 14-2	185
Answers to Challenge Problem 14-2	186
Determining the Rework Direction and Amount	187
Rectangular Method of Correction	190
Challenge Problem 14-3	192
Answers to Challenge Problem 14-3	196
Polar Method for Correction Amounts	196
Challenge Problem 14-4	198
Answers to Challenge Problem 14-4	198
Review of Rework	200

CHAPTER 15 Computing Geometric Tolerances for Designs — 202

Virtual Condition	204
Calculating Position Tolerance Using the Floating Fastener Tolerancing Method	207
Challenge Problem 15-1	208
Answers to Challenge Problem 15-1	209
Calculating Positional Tolerances Using the Fixed Fastener Principle	210
Challenge Problem 15-2	213
Answers to Challenge Problem 15-2	215

CHAPTER 16 Complex Tolerances — 216

Bonuses from Datums of Size	216
Calculating Double Position Tolerance Pertaining to a Single Feature	218
Double Bonus Applying to a Group of Features	220
Challenge Problem 16-1	222
Answers to Challenge Problem 16-1	224
Complex Rework	224
Challenge Problem 16-2	226
Answers and Hints to Challenge Problem 16-2	228

Unit IV
Geometric Theory

Unit Introduction — 232

CHAPTER 17 Geometric Element Analysis — 233

    Straightness—One More Time — 234
    Element Analysis — 236

CHAPTER 18 Surface Roughness Evaluation — 241

    Surface Roughness Analysis — 243
    A Final Word — 246

GLOSSARY — 249

APPENDIX — 253

    Challenge Problem — 253
    Rework Sheets and Formulas — 254

INDEX — 257

Preface

The subject of Geometrics is far more vital to manufacturing than a set of rules for dimensioning and tolerancing a part print. Geometric Dimensioning and Tolerancing concepts underlie the very basis of manufacturing. *If you are machining and inspecting any product, you are using geometric principles whether the design is geometric or not.* Geometric dimensioning and tolerancing is a systematic approach to controlling the entire manufacturing cycle from design to final inspection.

Without geometric working skills, you face many shop assignments with a handicap because you have the same challenges but less control of the process. It is not possible to make anything in classic manufacturing without using geometric concepts!

This is a book for people interpreting prints, and making, inspecting, and reworking machined parts; for programmers, toolmakers, and shop supervisors; and for any person requiring a deep understanding of the application of geometry in manufacturing.

Geometric designs start with and serve the function of the product. They provide the full natural tolerance in any mechanical assembly. Products they produce are easier to make and work better too. However, if the system is to work right, both the design and manufacturing areas must understand the working concepts—not just the definitions and drafting standards—and be able to manipulate and calculate. This book is about the skills that enable you to use geometric designs to their fullest advantage in the shop.

This book provides four sets of working skills that enable you to:

1. Read and Interpret geometric designs on prints.
2. Understand and Analyze the concepts of Geometric Control.
3. Calculate and Manipulate tolerances.
   Take full advantage of the system and use the rules.
   You will be able to inspect and compute geometric rework of potentially scrapped parts.
4. Think Geometrically
   Geometric concepts will make you a better problem solver with insight of the entire manufacturing process.

This deep understanding will bring about:

1. better inspection ability,
2. improved machine setups,

3. better quality machined and fabricated products,
4. more efficient CNC programs,
5. full use of CMM inspection equipment,
6. an overall perspective of manufacturing,
7. a background understanding of the dimensions and tolerances you find on prints, and
8. improved jigs and fixtures.

Geometrics are a must if quality and process are to be fully controlled. Geometric designs allow more specific control of part geometry. That is just what SPC needs to work best. Geometric designs focus close, custom control exactly where it is needed and loosen tolerances when ever possible. Properly used in the manufacturing process, they simply provide more usable control in any aspect of manufacturing, but especially in quality assurance.

The purposes of the four units are as follows:

— Unit I Foundation Skills. From facts and examples you will see why this is such an important skill in your shop training. We start with enthusiasm for the subject!
— Unit II The Working Concepts. One at a time, we will construct a working knowledge of the elements of geometric dimensioning and tolerancing and inspection.
— Unit III Application Skills. Here you can try out your new skills. You will manipulate and calculate real manufacturing challenge problems in machining, inspection, and rework.
— Unit IV Geometric Theory. We then extend your depth of understanding by looking at the theory behind the lessons and at present challenges and future developments. We end with a broadening of your insight.

An understanding of geometrics is complementary to your knowledge of Computer Aided Design (CAD), Computer Assisted Machining (CAM), Computer Numerical Control machining, and, especially, Coordinate Measuring for inspection. Geometrics is a bridge between each discipline. Today, with the assistance of computer manufacturing equipment, we can put geometric dimensioning and tolerancing to work in a far more practical way than ever before.

Geometrics and this book are about *Control* in manufacturing. You will be challenged, rewarded, and delighted at the dept of understanding you will gain. A working knowledge of geometrics is essential to your career now, and even more so in the future. As technology deepens, the need for geometric understanding will be ever more important.

Acknowledgements

The following people and organizations have made significant contributions to this book:

The Boeing Commercial Airplane Company, Training and Apprenticeship Division

Mr. Tom Clemmon—Instructor, Training for Industry, Everett, Washington

Mr. Ken Rouse—Instructor, Shoreline Community College, Seattle, Washington

Mr. Kris Snyder—Student, Sno-Isle Skills Center, Everett, Washington

Mr. Bill Coberley—Quest Tech, Salem, Oregon

Mr. David Genest and many others at Brown and Sharpe—Brown and Sharpe Corporation, Rhode Island

and, for the unit opener drawing Delmar Publishers Inc., from Jensen, *Geometric Dimensioning and Tolerancing for Engineering and Manufacturing Technology*, copyright 1992. Used with permission.

UNIT 1

Foundation Skills

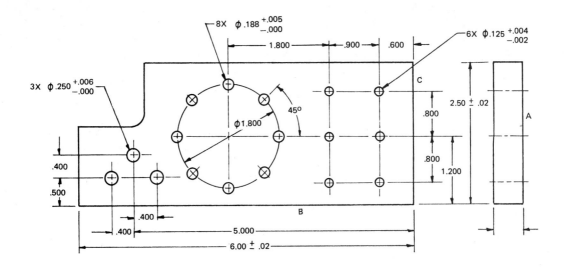

Unit Introduction

These are the building block skills you will need for two reasons:

1. To read and understand geometric drawings, and
2. To study the deeper, more technical aspects of actually applying geometrics in manufacturing.

This unit will provide a good foundation for all that you will learn later in this book.

CHAPTER 1

What is the Geometric Dimensioning and Tolerancing System?

A Tolerancing Control System

Geometrics is a system of control of the parts we make. Everything designed and manufactured has a functional requirement and an allowance away from that given requirement. Geometric dimensioning and tolerancing is a better way of controlling the entire process from design to manufacturing.

For example, in order for a pin to press fit into a hole, the bored hole would have a design requirement of .998-inch diameter with a plus .000 inch and a minus .002 inch allowable deviation (tolerance) from the basic size. The pin will press in and stay in but not be too tight. Also, the hole must be on a given location, and be round and perhaps perpendicular to some surface. Each of these design requirements must have some tolerance associated with the design on the print. The geometric system is a way of:

1. Calculating what the full natural tolerance can be, based upon the function of the part, which yields more usable tolerance on the print, and analyzing the design for the exact tolerancing possible.
2. Giving more specific control to individual features as required; loosening tolerances where possible and tightening where required. This results in improved quality and less scrap because the parts are easier to machine, thus they cost less.
3. Communicating this tolerance on the blueprint. Geometrics is mostly symbolic thus eliminating ambiguous and lengthy notes on the drawing.
4. Working with the tolerance in the shop. In the geometric system, if you know the rules, the given tolerance may be increased in two ways by the way you machine the parts. This flexibility of tolerances, made possible

only by using geometrics, is part of the expanded control you are about to learn.

All four factors above contribute to better product designs, less scrap, less cost, and more control of the process in general. The entire geometric system is based upon *Natural Function* of the design.

Example

Suppose that we are making an inspection block—a flat table upon which measuring instruments are to slide—out of cast iron. It is to be 4 inches thick.

Based upon the function of the part, what is critical about the design and machining; what requires close tolerancing and what does not? The most critical feature would be the flat top. Assume it is to be flat within .0002 inch. The top would need to be parallel to the bottom but this is not as critical as flatness because there are adjustable legs on the bottom. In terms of how it works (function), the four-inch thickness is not very critical at all.

Now, let's see what happens when we ignore these geometric requirements and use a nongeometric design to make the part. In Figure **1-1**, an oversimplified nongeometric print would accomplish the end result but the part would be costly, hard to make, and prone to scrap due to over-control of features. The designer controlled the flatness using the height tolerance.

As you can see in Figure 1-1, this print controls the flatness within tolerance but in so doing needlessly over-controls the parallelism to the base and the height too. All must be within the .0002 inch callout. In fact the bottom of the part must be machined as well and it too must be very flat to fit the requirements.

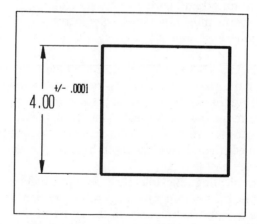

FIGURE 1-1 A nongeometric control.

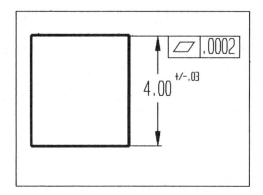

FIGURE 1-2 An improved geometric control.

Figure 1-1 simply doesn't fit the natural design requirements of the part. There is no individual control of specific design requirements. This example is extreme yet, without geometrics, designers are forced into similar but more subtle traps when dimensioning and tolerancing their work.

In Figure **1-2**, the print has been dimensioned geometrically. The geometric callout on the right is read "flat within .0002 inch." Now that the flatness is under control, the designer may assign a much bigger tolerance to the less important features such as height and parallelism. This part is much easier to make and less prone to scrap. Most important, however, is that it will work just fine at the lowest manufacturing cost. It would now be easy to add a parallelism control for the top of the part if so required or to refine the overall control of this part in any way. Each design requirement can be individually dealt with as required. And each tolerance would reflect the natural needs of the function of the design.

Statistical Process Control and Geometrics

Figures 1-1 and 1-2 can illustrate another important aspect of geometric designs: their importance to statistical process control (SPC). Briefly, SPC is a systematic approach to controlling the variation in the way we make parts. SPC allows manufacturers to clearly see where the variables are in the process and which ones cause variation in the product. SPC is a data collection tool that produces better products.

Geometric Designs Provide Individualized Control Points

A certain number of parts must be made before SPC can identify control points and variables. This is where geometrics comes into action. The part in Figure 1-1 has nearly no control points. To make this part per design would require a lot of extra machining to make it acceptable. However, the geometric part in Figure 1–2 has the controls separated by function. The height is separated from the flatness and parallelism to the bottom. Each function can be dealt with individually. The intervention points where the shop might be able to adjust the process are clearly defined and based upon the design function of the product.

Geometric designs work hand in hand with SPC; in fact, SPC only works right if the design is geometric. By not using geometric control, the designer simply handicaps the entire manufacturing process, including SPC.

Five Types of Feature Control

We will repeatedly use the word "feature." A feature is any part of a product that can be controlled using dimensions and tolerances. A hole, a tab, a surface, a thread, and a slot are all examples of features.

There are five large groupings of feature control that will be further analyzed in Unit II.

Geometrics define the tolerable limits of:
1. **Form**—This is straightness, roundness, flatness, and cylindricity.
2. **Profile**—This control is closely related to form but controls more complex shapes—profile of a line and profile of a surface.
3. **Orientation**—This control defines the tolerance of perpendicularity, angularity, and parallelism—the relationship of one feature of a part to another.
4. **Location**—This control deals with the position of the centerline of a feature. Concentricity is also included in this grouping.
5. **Runout**—This controls the surface wobble of rotating parts—a type of form control.

From this brief discussion you should remember that the Geometric System controls the dimensioning and tolerancing of the form/profile, orientation, location, and runout of part features on blueprints.

Control Zone—In all five groupings the end result is always the same. The print defines the design requirement and the geometric tolerance built around that requirement. This tolerance then creates a tolerance zone in which the feature or its centerline/plane must lie. We then must control either the surface of the part, as in the inspection table example, or a feature centerline (or centerplane), as in the bored hole example, of a part feature within the specified tolerance control zone.

As an example, consider the inspection table of Figure 1–2. Here the tolerance

CHAPTER 1 What is the Geometric Dimensioning and Tolerancing System?   □   7

control zone would resemble a sandwich. The surface of the part could deviate .0002 inch between the two limit planes. This control zone would float up or down to match the machined height of the part but must be able to fully contain the entire top surface of the part.

Any single line drawn on the surface of the part could not go out beyond these limit barriers and all lines must fall within the control zone. These single lines that you could draw on the surface of the part are called "elements." For the inspection table, the barriers would be a pair of perfect surfaces .0002 inch apart. Any line on the top of the table must not go beyond the tolerance control zone and *all* elements must stay within the control zone.

This is true for the entire subject of geometrics. For every print control, on every drawing, the print tolerance defines a perfect control zone built around the design dimension. Either the surface of the part or the centerline/plane must fall inside this control zone.

This is a constant in Geometric Dimensioning and Tolerancing (GDT). Either a surface or centerline/plane is controlled within a perfect tolerance zone. The shape of the zone will be different depending on the nature of the control but the principle remains the same. In the shop, you must make and measure your product to ensure that either the center-line-plane or surface elements fall within the control zone.

CHAPTER 2

What Advantages are There to Geometrics?

The Industrial Need For Geometrics

The Geometric Dimensioning and Tolerancing System didn't just happen. It was created or perhaps more correctly compiled by a group of military and civilian engineers, scientists, manufacturers, and teachers on a committee supported by the American Society of Mechanical Engineers. This group is called the American National Standards Institute (ANSI) Y14 committee.

This committee oversees geometric rules in this country and interfaces with the International Standards Organization for worldwide geometric standards. Much of this textbook is based upon the ANSI Y14.5-M standard and occasionally direct reference will be made to a certain section and number in Y14.5-M. It is advisable that you obtain a copy.

A Controlled/Supervised System

As new ideas, deeper understandings, or simplifications become possible, the ANSI committee incorporates them into the standards. This committee meets regularly and considers how to improve the system and keep it parallel with modern manufacturing.

From time to time, there are updates to the standards of which you must keep informed. In this book we will be learning the latest standards while also looking at previous procedures. In the shop, you will encounter older geometric designs that have not been updated and you must be able to interpret them to make products.

Example

The symbol for "symmetry" (centered about a centerline) was eliminated in the 1982 revision and thus the control was given to position. These changes not only streamlined the system but reduced the amount of printed data on the drawing. This aided CAD drawings because the drawing took up less space in the data base. Symbols require far less memory than groups of words. Changes such as this add to the universal understanding of geometrics and eliminate ambiguities.

It is anticipated that future updates will deal with changing technology. New ANSI standards are needed for more accuracy and a general mathematizing of the system.

Advantages to the Geometric System

The geometric system controls part geometry and improves the manufacturing process in nine different but related ways.

1. *Interchangeability.* The geometric system allows industry to create designs that function well and fit together in the best way, with the most natural tolerance. This is a prime directive of the geometric system.
2. *Functional Control and Gaging.* Because geometrics start with the function of the part as a basis for dimensioning and tolerancing, more custom tolerance is available. This process allows inspectors to check with functional gages. A functional gage simulates the extremes of fit—tightest and loosest. Functional gaging can save time in the shop.
3. *Full Advantage of Natural Tolerances.* Designs that are toleranced using geometrics will have more tolerance than prints that are dimensioned non-geometrically.
4. *Flexible Control of Designs.* Different aspects of a design require different degrees of control. Geometrics allow the designer to custom tailor the control. With geometrics, we fit the tolerance to the function. Without geometrics we fit the tolerance to our methods of tolerancing often resulting in false and more difficult controls than necessary.
5. *Flexible Control in the Shop.* You are about to learn that in geometrics a given tolerance may be increased using a process called "bonus" tolerance. There are natural relationships within functional fit that allow some flexibility of tolerances not found outside geometrics.
6. *Symbols Save Time.* Because the geometric system uses symbols to denote complicated instructions, both the drafter and the print reader save time. Complex concepts may be communicated with a few symbols and numbers. Even language barriers fall away using geometric symbols as long as the users understand the concepts behind the symbols.
7. *Universal System.* These symbols are understood by everyone involved and will be consistent from company to company, country to country.

8. *Controlled System.* The total system is controlled, supervised, and overseen by a national committee and will be updated to keep pace with industrial needs.
9. *Language of Analysis.* Thinking geometrically gives the user a deeper understanding of manufacturing. You are able to comprehend function and fit better. Without the concepts of geometrics it is very difficult to determine exactly what tolerance is possible within an assembly. Therefore, engineers and shop personnel are often guilty of tightening tolerances "just to be sure." This practice is costly and causes needless scrap due to unrealistic tolerances.

Symbols Used in Geometric Dimensioning and Tolerancing

Since geometrics is mostly a symbolic system, it is important that you know the symbols. Figure **2-1**, reprinted here from the ANSI Y14.5-M, is the symbols used in geometric dimensioning and tolerancing. It is not required nor recommended that you learn these symbols at this time but it would be wise to scan them and begin the process.

A more complete study of the symbols will follow. Scan them now and return to this page often. If you wish to be proficient in geometrics, you must be able to interpret the symbols.

Tolerance Versus Cost

A final word of why you should learn geometrics. This is a critical fact that all manufacturers should understand. By correctly using geometrics, costs can be reduced tremendously. By correctly using a combination of statistical process control and geometrics, products get the best of both—they cost less and have improved quality.

In Figure **2-2** (page 12), is an eye opening fact sometimes overlooked by designers and shop managers: the tighter (smaller) the tolerance gets, the higher the manufacturing cost gets. We are going to look at cost versus tolerance.

Example

Suppose we are machining the inspection table previously discussed and we wish to track the cost per unit and see how tighter tolerances affect the cost. We will only consider the flatness tolerance for the top. Initially, for this example, let's start with a tolerance of +/- .002 inch for flatness. This might be possible on a standard milling machine using a fly cutter. The estimated cost for this example would be $100 per unit.

CHAPTER 2 What Advantages are There to Geometrics? □ 11

SYMBOL FOR:	ANSI Y14.5	ISO
STRAIGHTNESS	—	—
FLATNESS	▱	▱
CIRCULARITY	○	○
CYLINDRICITY	⌭	⌭
PROFILE OF A LINE	⌒	⌒
PROFILE OF A SURFACE	⌓	⌓
ALL AROUND–PROFILE	⌭	NONE
ANGULARITY	∠	∠
PERPENDICULARITY	⊥	⊥
PARALLELISM	∥	∥
POSITION	⌖	⌖
CONCENTRICITY/COAXIALITY	◎	◎
SYMMETRY	NONE	≡
CIRCULAR RUNOUT	*↗	↗
TOTAL RUNOUT	*↗↗	↗↗
AT MAXIMUM MATERIAL CONDITION	Ⓜ	Ⓜ
AT LEAST MATERIAL CONDITION	Ⓛ	NONE
REGARDLESS OF FEATURE SIZE	Ⓢ	NONE
PROJECTED TOLERANCE ZONE	Ⓟ	Ⓟ
DIAMETER	⌀	⌀
BASIC DIMENSION	⎡50⎤	⎡50⎤
REFERENCE DIMENSION	(50)	(50)
DATUM FEATURE	-A-	⌴ OR ⌴—A
DATUM TARGET	⌀6/A1	⌀6/A1
TARGET POINT	×	×

*MAY BE FILLED IN

Figure 2–1 Comparison of symbols (reprinted with permission from ANSI Y14.5-M. 1982.

Now let's decrease the tolerance to +/- .001". We needn't change much in our process other than slowing down a bit. The cost won't change very much, let's say to $150 per unit.

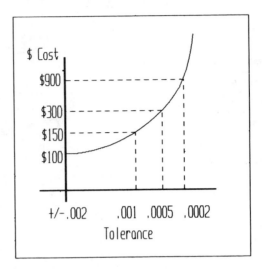

Figure 2-2 Cost versus tolerance. The effect of tighter tolerances on manufacturing cost.

Now put a tolerance of +/- .0005". This might require grinding and, at the very least, a roughing and finishing operation. The costs are going to start climbing quite fast here to $300 per unit. Additionally, there is going to be a better chance of scrap parts, more inspection and handling of the part, and far more time needed to machine to completion.

Now, at +/- .0002, it will require extra operations and a lot of handling of the part. Perhaps some stress relieving of the casting after machining too? At any rate the cost is probably going to more than triple—to $600 per unit. As you can see from this example, the cost is growing at an exponential rate.

If we cut the tolerance in half, we don't simply double the cost but, due to extra operations, it might grow by as much as ten times. As a final argument, let's put on a tolerance of +/- .00000001 inch. Could we machine this ridiculous requirement? If you had millions of dollars, you still couldn't machine this part. Every part is going to be scrap! At some point, tightening the tolerance sends the cost out of reason and all parts become scrap due to the difficulty of making them.

The lesson to be learned here is to take advantage of the tolerances that designers build into geometric prints. In their efforts to turn out quality work, craft people must be careful to not over-control their "in-house" statistical process tolerance.

Manufacturers must balance quality against cost. The objective is always to produce products with minimum variation from the design requirements. However, in so doing, a very important question must always be asked: "Is the SPC needlessly driving the product cost up?" Or is the tighter control costing nearly the same but producing a better product?

UNIT II

The Working Geometric Concepts

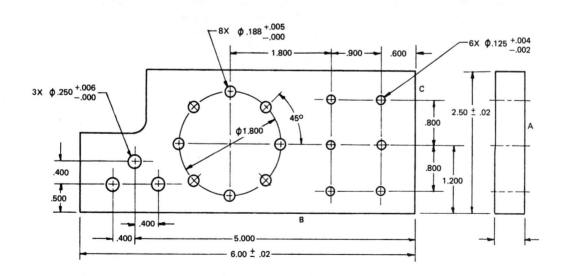

Unit Introduction

The overall goal of this book is to teach you a full working ability in geometrics. Here in Unit II you will learn the building block skills of interpreting drawing symbols and to understand the individual geometric concepts. You will also learn how to inspect per geometric concepts. With these concepts mastered, you will learn tolerance calculation and manipulation skills in Unit III.

To reach this working level of competency, you need the building blocks—a working knowledge of the individual geometric concepts. We will examine each geometric concept. We will see the similarities and differences in the thirteen different feature controls within the system.

Some are simple and some complex. Make sure you grasp each concept before moving forward. Discuss these concepts with fellow students, instructors, engineers, and journey level machinists if you need help. In Unit II we are building an entire system. The real challenging work will be possible once you have the full range of skills.

Since this book is an application/doing book based upon the reference book ANSI Y14.5-M, you should have a copy of Y14.5-M available as a reference while studying. Reference will be made occasionally to Y14.5-M for two reasons: first, to familiarize you with this important document—you will be referring to it in your career, especially as new issues and revisions arise. ANSI Y14.5-M is a living document. Second, you will be referred to ANSI to expand upon the explanation of some finite detail which is too tightly focused for the scope of this book while you are learning the entire system, but one to which you will need to return later in actual industrial practice. You need not concentrate on the ANSI explanation at this time, but simply scan through it and know where to find it. Know that a further explanation does exist and be ready to use it in real application, after mastering this book.

CHAPTER 3

Datums—A Basis for Measuring and Position

Concept 1: There are twenty-eight major concepts that compose the geometric system. Our first, the concept of Datums, is important to learn because datums form the foundation of geometric dimensioning and tolerancing (GDT).

A Starting Point for Dimensioning, Measuring, and Positioning

There are two reasons datums are important in geometrics:

1. Datums eliminate guesswork on measurement. Datums provide an exact basis for measurement, a reference from which the process of measuring, machining, and inspection is started.
2. Datums control the relationships of features. They provide a basis for controlling characteristics such as parallelism and squareness.

Understanding Datums, Example 1.

In order to understand the concept of Datum, answer the following question. Figure 3–1—How tall and how parallel is the top of this exaggerated part compared to the bottom?

It is difficult to respond because you do not know from where to take the measurement or reference. Using a micrometer, you would get a different answer depending on where the micrometer tips were located because there are irregularities in both the top and bottom surfaces of the part. Without a datum, depending on where you measure you could have many answers. Which is right? A "Reference Surface" is needed.

16 ◻ Unit II The Working Geometric Concepts

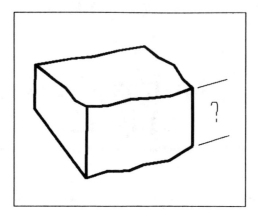

Figure 3–1 How tall and parallel is this part? The answer is not precise.

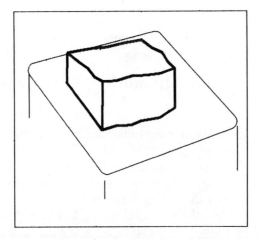

Figure 3–2 On this inspection table we can answer the question. The table becomes the reference datum. Now the limits of height and parallelism can be determined.

To fully appreciate this example, keep in mind that "functional fit" is the basis for all of geometrics. The main purpose of geometrics is to better serve how manufactured parts assemble. Since the height of the part is in question, the functional fit would be the minimum space into which this part would fit. You could answer the question of how tall this part is if you somehow could determine the Minimum Height of a space into which the block would fit. To answer the question, we need to establish a datum reference plane.

Now, set the part on a flat surface such as an inspection table, then test the height and parallelism.

In Figure 3-2 it is now possible to measure the part. The inspection table becomes DATUM -A- as established by the features of the part—in this case, the three (or more) contact points between the part and table. Now, you can easily sweep an indicator or the scribe of a height gage over the top to determine the height of the part and also the parallelism to the base datum. You are comparing the top of the part to Datum -A- established by the part features.

The datum deals with the bottom irregularities and makes it possible to answer the question of how tall and how parallel the part is. You are really determining into what space this part will functionally fit and how parallel the top of the part is compared to Datum -A-. Datum -A- then represents a mating part or the functional limit of the bottom surface of the part.

Datum Features Establish Datums

A Feature (a hole, a flat, a centerline or curved surface, a thread, a boss, a slot, or a diameter) is any part of a design that can be dimensioned and toleranced. All features have a theoretical perfect shape and actual irregularities of the shape. The datum then deals with those irregularities.

When a feature is used to identify a datum, it is called a "Datum Feature" in geometrics. As in Figure 3-1, the part features (the bottom of the block) established the datum. The datum is considered perfect while the datum feature(s) that establish the datum are not perfect. Notice that the datum was not on the part but rather a perfect surface touching the part. In this case the datum was the inspection table.

So far we have learned: A datum is a theoretically perfect surface or line established by the part features.

In the shop, for an inspection gage of this part, Figure 3-1, you could make a precision ground block that simulates the smallest mating space available per the design. It would represent the tightest space into which the block in question must fit including the mating parts size tolerance. If this functional gage slips over the part, then it will assemble. You would also need a second gage, or a step in the first, to determine if the tested block was too short as well. This is the concept of "functional gaging"; to simulate the extremes of fit per the design.

Datum, Example 2.

In Figure 3-3 you need to determine how far the hole with its exaggerated irregularities is from the edge of the part. Think about functional fit and the concept of datums to determine the answer before reading the next paragraph. How would you inspect it?

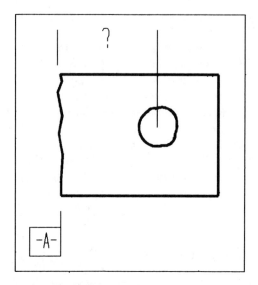

Figure 3–3 To measure this distance requires two datums.

The answer lies in establishing Datum -A- and also in where the functional hole is located. First, lay the part on an inspection table against datum surface feature A to establish the reference (Datum -A-). Next, establish where the geometric hole is actually located. The functional hole really means the diameter and location of the largest cylindrical space established by the hole—the largest cylindrical space available.

How? A test pin of the largest possible size that would fit into the hole could do the job. This would be the true cylindrical space available after we take into account the irregularities of the hole. The distance from the center of the pin to Datum -A- is this part's functional fit—also known as it's "virtual condition."

"Virtual Condition" is the sum effect of all the toleranced factors—size and shape, orientation (is the hole parallel to the edge), and location—on the final total result in terms of fit. The virtual condition of the real part would, in this case, be the largest available cylindrical space that is parallel to Datum -A-.

The distance from the center of the test pin to the edge datum would solve the problem and would also determine the actual functional fit of this part—that is, the functional distance of the cylindrical part of the hole to the edge Datum -A-. Understanding this illustration is important to your knowledge of datums!

Can you see how a functional gage might be made to inspect this part? It would simulate the worst case—the test pin must simulate all the possible space that the mating part could occupy (biggest possible pin with maximum misloca-

CHAPTER 3 Datums—A Basis for Measuring and Position □ 19

tion) within the design tolerance. If the gage can assemble with the part in question, the part is correct.

Datum Callout Symbols and Order of Importance

Datum Callout

A datum is specified by the placement of a box around a letter called a "datum identifier." A dash will precede and follow the datum letter. Datums will either be surfaces or centerlines.

Datums That Are Established by Part Surfaces

A leader or extension line will then connect the identification box to the surface of the part. If the datum is from a part surface, the datum identification box will be drawn in a view in which the datum surface appears as a line (see Figure 3-3).

Datums That Are Centerlines

Sometimes a datum is an axis or centerplane of an object such as in Figure 3-4. If the axis (not the surface) of a part feature is to be used as a datum then the identifying box will be placed adjacent to the *size* information for that feature and not be connected to the surface of the part.

These datums are easy to discuss but often present a real problem in inspection.

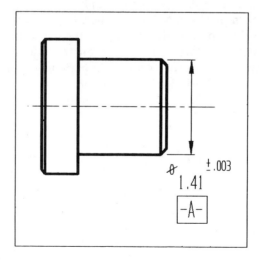

Figure 3-4 A center line datum identifier.

The problem is that reference is being made to a centerline or sometimes a centerplane that only exists as a composite result of the entire outside shape of the part.

In the example, the centerline datum can be envisioned as the true center of the smallest diameter perfect cylinder that would contain the part. The datum would just contain the part with its irregularities. In practice, to establish Datum -A- here, we would need to confine the part in a precision collet or some other precision holding devise. The center of the holding devise would then be Datum -A-.

Datum Priorities Per Design Function

On a design, the most important datum in terms of function is the primary datum for the entire part which is often called DATUM -A-. -B- would be next in importance and -C- would be third and so on. Other letters may be used and there may be any number of datums on a drawing. In fact any feature that is used as a reference for another feature is either a formal assigned datum or must be *treated* as a datum to measure or machine the part. This informal datum is called an "implied datum." Any feature of the part may be a datum.

Individual Feature Control Priorities

Datums are chosen by their importance to the entire part or assembly. Each feature of the design will then relate to one or more of these datums. When you are making a part, you must pay close attention to them and their order of importance. The datum reference for each specific control will be prioritized as to the combination and order of datums that apply.

Take, for example, the location of a single hole in a part, Figure 3–5. This control sets priorities for the process through which the part is to be machined. The datum reference letters may be found in any order but the order in which they appear in the control frame will indicate their order of importance to the design function for that feature. Even though the design has many datums, those in the control frame are the datums that pertain to this specific feature control callout.

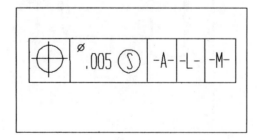

Figure 3–5 The order of appearance sets the priority of importance for this control.

Suppose we find this geometric location callout on a print such as the one found in Figure 3–5. This means that to machine or inspect the location of this hole you must make every effort to first hold the part against a locating (vise, chuck, etc.) fixture on Datum surface -A-. Next establish -L-. Then, with these under control, establish datum feature -M-. You could change the inspection results if you treat the datums differently (example to follow).

When we study datum frames in this chapter you will see how the control has specified the primary, secondary, and tertiary datum order for this specific control callout—the location of the hole. For control of this feature, Datum -A- is primary, Datum -L- is secondary, and Datum -M- is tertiary.

Inspection Error Example

To understand datum priorities consider Figure 3–6. The design callout controls the perpendicularity of Datum Feature B (the bottom edge) to Datum -A-. During in-

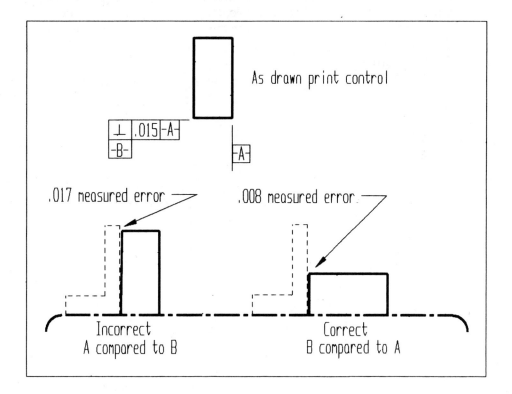

Figure 3–6 The orientation of Side B is controlled to Side A. The incorrect datum priority yields a scrap part when in fact it is acceptable.

spection, this requires you to compare Datum Feature B to Datum -A-. It is to be square within .015 inch. Suppose that the controlled corner is less than 90 degrees.

Standing the part up as found on the print, then checking the perpendicularity with a square, is wrong. Incorrectly measured, the example error would be say, .017 inch—this is unacceptable per drawing. Correctly laying the part on Datum Feature A, and comparing side B, you would find far less error. Can you see why mis-prioritizing of the datums caused you to reject the part when it was actually within tolerance? This is a simple example of subtle, more complex traps that can occur if you do not pay attention to datum priorities during machining and inspection. With holding or inspection tools, you must hold the part on the highest priority datum first, then establish the secondary then the tertiary.

Datum Letters Used

Any letter of the alphabet may be used to identify a datum except I, O, and Q. Occasionally you will encounter drawings that have assigned single letters through the alphabet and then started with double letters such as -AA-, -BB-, etc. Nearly any dimensioned and toleranced feature on the drawing may be assigned as a datum.

Compound Datums

You might encounter a datum symbol such as R-S within a single box (see Figure 3–7). This means that the datum is made of (compounded) two datums. In Figure

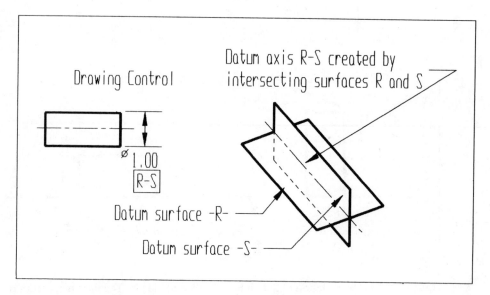

Figure 3–7 A combined axis datum symbol.

3-7 it is the line formed by the intersection of two datum planes. This is a datum method with which the designer could control radial features on round parts such as a bolt circle drilled around the rim of this part. The part illustrated in Figure 3-7 is referenced by its centerline which is actually line R-S. Planes R and S then would provide some control of radial features. This then would provide a basis for measuring the relationship of the keyway in the center to the holes on the rim. There are other ways in which datums may be combined into a single reference.

The second meaning of the symbol R-S is that the datum is formed by contact with two surfaces of the part at the same time. For more information, see ANSI Y14.5-M 4.4.5.1.

Definition

— Datums are theoretically exact planes, surfaces, or lines from which reference may be taken for measurement or relationships. Datums are established by datum features on the part.
— Datums are established by either point or surface contact with part features.
— Datums simulate the assembly world outside the part being inspected.

They are not the part, but established by the part. Datums give the user a method of dealing with the actual irregularities of the part. Datums provide an exact basis for measuring and positioning.

Datum Target Symbols

Specific Points to Establish Datums

The entire surface of a part may not be reliable enough to establish a datum, as in the case of a rough casting or forging, or the designer may simply wish to specify exactly where and how the part is contacted to establish a datum. Then the designer will call out a specific set of points on the drawing that establish the datum. These are called "Datum Targets" and they are indicated by datum target symbols.

These target symbols denote where and how the part is to be contacted in actual machining or inspection. In the machine shop, fixtures are built based upon the datum target points on the drawing. The fixture will have tooling points that contact the part in the specified locations by the datum target symbols. So datum target symbols become "Tooling Points" on actual hardware. The contact may be a point, line, or small surface area—usually a circle.

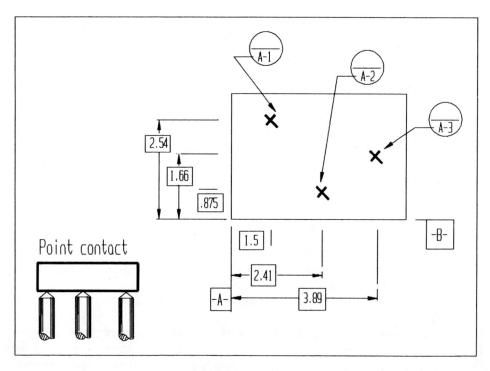

Figure 3–8A Points identified by datum targets and basic dimensions.

The placement of the target on the actual part—that is, the location and size of the area in which the designer wishes the part to be contacted—has three forms. In all forms, the symbol on the drawing is a circle.

Example 3–8–A, Point Contact (Figure 3–8A).

The symbols identify the datum and the point of contact. A-1, then, denotes the first point needed to establish surface A. The physical location of the contact on the part, the bold X, is dimensioned from part feature datums.

In practice, the point will have some size; i.e., tooling points will have a contact radius or small flat on the tip of the contact pins.

Example 3–8–B, Area Contact (Figure 3–8B).

The symbol now denotes the size of the target zone for contact. The part must be contacted within an area of 1.5 inches in diameter at each location. The location of the center of the shaded target zone is dimensioned from part feature datums.

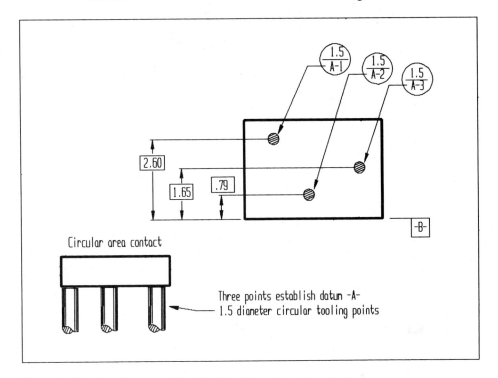

Figure 3–8B Circular areas identified by datum target symbols.

Line Contact

A third form is that of line contact which is not illustrated. Here the tooling is the side of a round pin or a ground straight edge.

No Targets Shown

If a surface is designated as a datum on the drawing but no target symbols exist on the drawing, then the location and method of contact is optional. If there are no datum targets on the drawing, the shop may choose to use a complete surface as a locator. Examples might be in a vise, angle plate, collet chuck, or any standard shop holding or inspection fixtures. The choice of a shop-made fixture using tooling points is also possible.

Temporary Datums

Due to complex part shapes or difficult machining sequences, a designer may specify a temporary datum. Temporary datums are used to get machining started, to establish another more reliable datum surface elsewhere on the part. Castings, weldments, and forgings may have small protrusions that have only one purpose, the establishment of a temporary datum. They are usually machined off the part at a later time.

Datum Frames

Datum Frames Define a Part In Space

The next step in using datums is when they are linked together on the design in a complete set called a "Datum Frame." A datum frame is complete when it defines the part in three dimensional space. A manufacturing tool (mill, lathe, or inspection fixture) must then simulate the design datum frame in such a way as to not allow the part to move in any direction and to hold the part against primary, secondary, and usually tertiary tooling points.

The formal definition of datum frame starts from the Fundamental Datum Frame. The fundamental datum frame comprises three datums at 90 degrees to each other. This fundamental datum frame provides a three dimensional starting point for measuring and machining. The fundamental datum frame would always work if we manufactured parts that were shaped like a brick. We need more custom tailored frames in real manufacturing.

To deal with design shapes that are not bricklike, a Secondary Datum Frame is constructed that fixes the location of the part in all directions possible and provides a basis for measuring and machining feature relationships.

An example of a secondary datum frame is a fixture that holds the part for machining. It is constructed around the shape of the part design. It locates the part using tooling points that contact the part at the specified Datum Target Points if they are assigned.

Datum Frame Example.

Datum -A- (Figure 3-9)—To understand the concept of a datum frame, hold a book with your hand. The Primary datum is established by three contact points. This is usually the most important feature of the part or the most reliable for initial reference for machine operations. Often datum -A- is the most critical surface or axis of the part in terms of function.

Note: to establish a flat plane, three points (not in a straight line) are required. Try this yourself: try to hold a book with just two finger tips. The book rocks along the line between the finger tips. Now, add a third finger tip and notice that the book cannot rock at all. But it can slide in all directions along the plane just established by your finger tips.

CHAPTER 3 Datums—A Basis for Measuring and Position ☐ 27

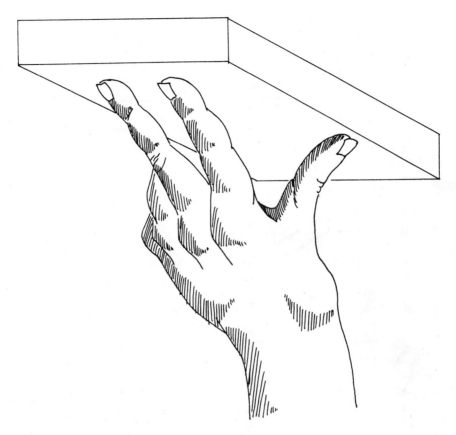

Figure 3-9 Three support points establish a primary datum.

Datum B (Figure 3-10)—The Secondary datum -B- is established by two points. This is the second most important or reliable feature. Since the part can now only move in a single plane, only two points are required to stop the part from rotating on plane A. Again, you can simulate this with your finger tips and a book. Once the two secondary points are added, note that the part can only slide in a straight line.

Datum C—The Tertiary datum -C- is established by one point only. The third point stops the part from sliding in a straight line and completes the containment of the part—it cannot move at all when clamped into the frame.

In the datum frame example above, the part being contained was a solid rectangle; therefore the datum frame had to have three mutually perpendicular datums. This gave us a convenient starting point for dimensioning and machining this part. But what about real non-rectangular parts?

28 □ Unit II The Working Geometric Concepts

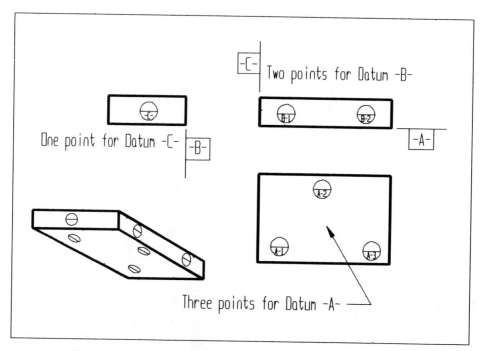

Figure 3–10 A complete datum frame.

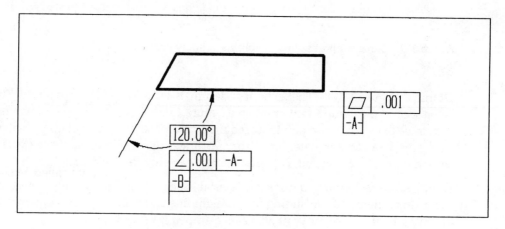

Figure 3–11 An angular relationship between datums -A- and -B-.

Not all datum frames are so simple. The purpose of a datum frame is to define and contain a part in space, to give a starting point for all measurement and relationships. The design shape of the part will establish the shape of the datum frame, which can be any shape as long as it holds (confines and defines) the part in space. In theory, we start with the primary frame—three 90 degree datums—then build further datums from this basic original system. The constructed secondary frame then has reference back to the original three perpendicular datums.

In actual shop practice, to make holding/inspection tooling, the machine/inspection environment is the primary frame. We start with a single datum—the most important surface or feature in the design—then build the successive datums from that starting point. In Figure 3-11 we would first establish surface A upon some type of datum -A- locator. This would be a ground plate or a set of tooling points that contact the part at datum target points. Next, datums -B- and -C- would be built with the correct relationship to the fixture Datum -A-.

Control of Datum Features and the Relationship Between Datums

In Figure 3-11, the part is not square. This requires some control of the datum relationships. When these irregular datum frames are set forth on the drawing, the relationships between the datums is controlled geometrically. The angle of datum -B- to datum -A- is controlled in Figure 3-11.

Also, datum features themselves may be controlled. Notice in Figure 3-11 that the designer required the surface of the part that established datum -A- to be flat within .001 inch. A flatness control would then lessen the irregularities on the part.

Datums of Size

Any part feature may be assigned as a datum and often the assigned feature may have some tolerance for size itself—thus the datum may be bigger or smaller. That is called a "Datum of Size." In Figure 3-12, the hole can vary between .745 diameter and .755 inch. Note that the 1.00 inch slot is related to the .75 diameter hole. Since the hole is the basis for the slot, it must be treated as a datum. The hole has tolerance. We will see later how as this hole varies from a state called Maximum Material Condition (MMC), the position tolerance for the related feature might grow through a process called "bonus tolerance."

In this case, the MMC diameter size would be .745 inch. If the actual hole was other than .745" the position tolerance could grow directly. We will study this in depth later but for now you must simply understand that there are *datums of size*.

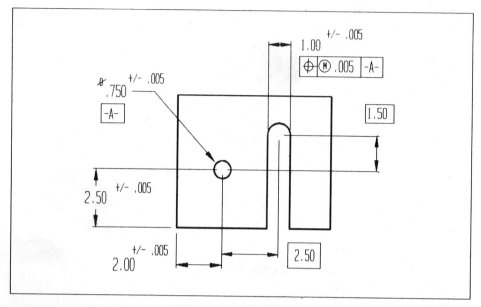

Figure 3–12 A .75 diameter hole is chosen as a datum.

Implied Datums

In Figure 3–12, suppose the designer did not assign a datum to the hole. You can see from what you have learned thus far that to measure the 2.5-inch callout, you must treat the .75 diameter hole as a datum even though it is not called out. This becomes an "implied datum."

When relating one feature to another, such as the slot to the hole, we must treat the basis feature (the hole) as a datum whether it is assigned or not. If a feature is used as a basis for the location or orientation of another feature, then the basis feature becomes an implied datum (if not assigned as a formal datum on the design).

Challenge Problem 3–1

The objective of the following activities is to get you started searching prints for datums and related geometric features. They are all based upon appendix drawing—Stabilizer Bracket. To check your understanding, the answers follow directly after the problems.

1. Identify the individual features that establish datums on the print. Overscore these using a transparent marker. How many do you find in the design?

2. What part feature establishes datum -C- ?
3. Which datum is a cylinder?
4. How many datums are flat surfaces?
5. The 1.190-, 2.502-, and 3.507-inch dimensions all refer to which datums?

Answers to Challenge Problem 3-1

1. There are three.
    Datum -A-—the bottom of the part
        -B-—the round boss
        -C-—the inner face of the wing
2. The inner surface of the left-most wing.
3. Datum -B-, the .500 boss that protrudes out the front view.
4. Two datums. -A-, the bottom of the mounting pads
        -C-, the surface of the left wing
5. They refer to datum A. To machine or inspect these dimensions, the part must first be held against the pads that form datum feature -A-.

Review of the Concept of Datums

Definition

Datums are theoretically exact planes, surfaces, or lines from which reference may be taken for measurement or relationships. Datums are established by datum features on the part.

— Datums are a starting point for dimensioning, measuring, and positioning.
— Datums provide an exact basis for measurement.
— Datums provide a basis for controlling relationships such as parallelism and squareness.
— Datums deal with real part irregularities.
— Datums are theoretically perfect.
— Datums are established on real parts by point, line, or surface contact.
— If the contact points are specified on the design, they are called datum points. In the shop tooling they are called tooling points.
— Datums allow the manufacturing of functional gages.
— Datums may be based upon part surfaces or center lines.

- Datums have design priorities based upon part function.
- Any part feature may be a datum and may be controlled.
- A datum may have size and tolerance on that size.
- Datums may be comprised of two other datums—compounded.
- A complete datum frame confines and defines the part in space. It provides a basis for measuring and machining.
- A primary datum frame is comprised of three mutually perpendicular planes.
- There are primary, secondary, and tertiary datums based upon the design.
- A secondary datum frame is built around the design shape of the part.
- When any control or dimension refers to a feature of the part, the feature must be treated as a datum. If the feature isn't assigned as a formal datum it becomes an implied datum.
- Datums simulate the assembly world around the part.

CHAPTER 4

Interpreting Feature Control Frames

Concept #2: Nearly all the information conveyed on a geometric print will be found inside a box called a "Feature Control Frame." These frames tell the user what type of control, how much tolerance, which datums are to be used for reference, and other modifying information such as maximum material condition (MMC) which we will study later.

Much of the advantage of geometrics lies in the ease with which a designer or drafter may communicate controls and relationships of features with symbols only. The print reader must be able to correctly interpret these symbols and their meaning. The most common location for a feature control frame is adjacent to the area affected by the control. This may also be called a "Feature Control Symbol."

Example of Feature Control Frames

Feature control frames are found in three forms on prints:

1. Single
2. Combined
3. Multiple

Single Example

The example in Figure 3–11 shows a single control on the left side of the print. Datum -B- must be at 120 degrees to datum -A- within a tolerance of .001 inch. Another single control is found to control the flatness of datum -A-. Note too that

34 ☐ Unit II The Working Geometric Concepts

a surface or other feature may be controlled then designated as a datum as well. In Figure 3-11, notice that the angular surface is datum -B-.

Combined Example

A drafter or designer may choose to combine two controls as in Figure 4-1 where you are told that the surface is to be perpendicular to datum -A- and also to be parallel to datum -B-. Both have a geometric tolerance of .002 inch.

Multiple Example

Feature control frames may be combined to indicate differential control tolerances. In Figure 4-2, top right, two control frames are listed to control perpendicularity and parallelism to datum features -A- and -B-.

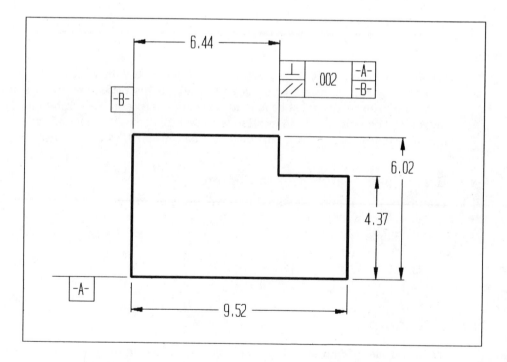

Figure 4-1 A combined feature control frame.

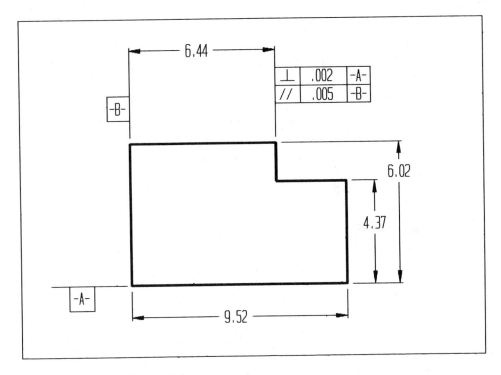

Figure 4-2 Multiple control frames.

Information Found in Frames

Figure 4-3 is a generic control frame. Here we are concerned with acquainting you with the categories and combinations of information that may be found within the frame—not the specifics. Not every category found here will appear in a specific control frame in a design. In fact, many frames may simply be a form symbol and a tolerance with nothing else (see Figure 3-11, right side).

Information will appear in the order of the example but not all categories will be found in every control:

- The Control Symbol—One of 13 possible (14 previous to 1982)
- Tolerance Zone Shape—Diameter symbol for round zones. No symbol—zone is parallel lines or surfaces
- Actual Control Tolerance
- Applicable Material Condition Modifiers—MMC, LMC, and RFS
- Datums to which the control applies—in order of priority
- Material Condition Modifiers may also be linked to datums

36 □ Unit II The Working Geometric Concepts

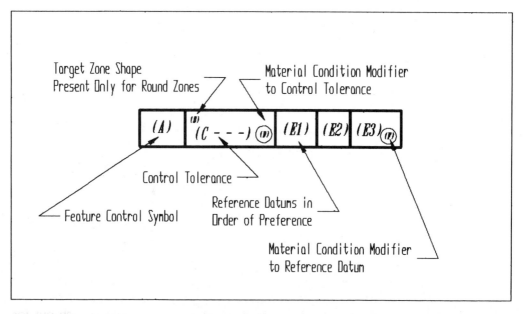

Figure 4-3 The order in which information appears in a feature control frame.

Scanning a Print

Often a drawing is complex with vital information scattered over the entire drawing. When initially reading such a print, it is easy to miss important notes and geometric symbols. Below is a suggested procedure to pick up all the details you must have.

1. Scan the front view and relate it to the other views. Identify the general shape of the part. Examples: round, donut, rectangular, and so on.
2. Locate three major features in the front view and relate these to other views. Take advantage of the orthographic projection principles used to project views out of the front view. Use a straight edge to project lines and points from one view to another.
3. At this point it is a natural practice to name the object's similarity to some familiar object. Examples: trumpet, hammer, fork, and so on. This gives all those working on the part a quick visualization of the object.
4. Return to Step 2 above and locate smaller and smaller details and features from the front view to each related view. Again, use the orthographic projection method of locating details in adjacent views.
5. At this point you are ready to locate notes and controls that have application to your work. You now have a grasp of the object's shape.

First scan the entire drawing for the obvious geometric control symbols. Next, either visually or using a pair of straight edges, divide the drawing into sectors—scan each sector more carefully. Most commercial drawings have a letter-number grid upon the margins that is useful in this search.

Often, a print copy may be available for layout, programming, or inspection purposes. This drawing is kept away from the normally revised drawings. Use caution on these drawings as they may not be up to date. However, you might be allowed to highlight geometric controls and other pertinent information as it is found using an overscoring pen.

Caution: *This may be against shop policy.* Many shops have a strict policy about writing on a print. Check before ever writing on a design print.

Also be cautious in using these utility drawings. They are not usually updated. These drawings must be kept away from the work area. Watch out for revision level problems.

Challenge Problem 4–1

Answers follow

Refer to appendix drawing—Stabilizer Bracket and Figure 2-1—the ANSI Control Symbols
1. Outline or overscore the feature control frames found on the print. How many did you find?
2. Which control frames denote combined controls?
3. How many are multiple controls?
4. Note that some controls refer to datums and some do not. Identify the controls that require datum reference.
5. Later in this text, we will study a great deal more why some controls require datum reference while others do not. However, in preparation for that study:
    A) Can you explain what the common basis is for all the controls that do require datum reference?
    B) Note the common grouping for the controls that do not. What is it?

Answers to Challenge Problem 4–1

1. There are 12 individual feature control frames. Note—the boxes around individual dimensions are not feature control frames. These boxes note that the dimension is a perfect target number. We will discuss this concept later in detail.
2. None are combined controls.
3. One—The control of flatness within .002 inch and the control of parallelism within .004 inch with reference to datum -C-.
4. Nine of the controls refer to datums.
   Four control *Position* features with reference to datums
      (one of these position controls relates to three features).
   Three control *Parallelism* of features.
   One controls *Angularity* of a feature.
   One controls *Perpendicularity* of a feature.
5. (A) All the controls that refer to a datum are either positioning or describing the orientation of a part feature with respect to another part feature. The feature that is used as a basis reference therefore must be a datum since this is a formal geometric design.
   (B) The controls that do not require datum reference are all controls of *form*.

CHAPTER 5

Geometric Tolerance Zones—the Control Tool

In the remainder of Unit II, we will learn the individual characteristic controls—what they are, how they differ, and how they are similar. You should understand that the entire system of geometric dimensioning and tolerancing is linked together. It truly is systematic—it all makes sense in its approach to control of part geometry, dimensioning, and tolerancing. You will be much more effective in manufacturing work if you envision this subject as a system of connected principles and concepts. The final goal of calculation and manipulation of the system requires an understanding of the interconnected principles.

> **Concept 3: The Tolerance Zone:** The major connection between all geometrics controls is how the different controls define the tolerance zone. All of the 13 geometric feature controls first define a perfect model around which a theoretical tolerance zone is constructed.

What Entities Are Found In The Tolerance Zone

Once the tolerance zone is defined, the part must be machined in such a way that one of two possible entities must then fall within the zone for the part to be within geometric tolerance. Which entity depends on the nature of the control. For example, flatness would require control of surface elements while concentricity controls the centerline of a feature.

Either the part *surface elements* or a *feature center* (centerline or centerplane) must fall within the zone for the part to be within geometric tolerance.

Two- and Three-dimensional Surface Controls

All geometric controls that deal with the outside surface of designed parts are either two dimensional, such as straightness and circular runout, or three dimensional, such as flatness and total runout.

For example, a straightness control is defined as a pair of model tolerance lines built around a perfectly straight line (Figure 5–1). The distance across these theoretical tolerance lines is the control tolerance limit. Any single element running in the control direction must fit into the tolerance zone. For the part to be acceptable per design, every element on the surface must be able to fit within an individual tolerance zone. Each element is an individual test. If they all pass, the part is straight within tolerance. This is called a single element or sometimes an individual element control. Individual element controls are two dimensional.

Now, if we wished to control flatness, the model would be a perfectly flat plane around which a tolerance zone would be built. This would then require all elements on the surface to fit into the zone.

This is the difference between two and three dimensional controls: straightness is a single element control while flatness controls all elements on the surface. Inside the tolerance zones, elements can deviate in many ways as long as they do not leave the defined tolerance zone.

Can you see the 2D/3D similarity between roundness and cylindricity controls?

> *Remember: Elements Are Lines on the Surface of a Part Feature.* You could draw an element on a part using a pencil. It is a line running in the direction of the control.

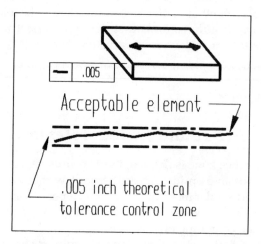

Figure 5–1 An element is a line that could be drawn on the surface of a part.

A roundness element would therefore be a circle perpendicular to the part axis, along the controlled surface. The objective of geometric surface controls is to define what the surface shape and/or the relationship of a perfect element is (straight, round, parallel, etc.) and then define the size and shape of the tolerance zone built around that perfect model, in which real elements must lie.

There are an infinite number of elements on any surface. To test the part shown in Figure 5-1 for straightness, an inspector might make five checks across the top surface. Each test would run in the direction of the test element determined from the drawing. The element appears as a straight line in the front view thus it extends right and left across the top of the part. That is the direction in which the part must be straight. Once a sufficient number of elements have been found straight, you could assume that the entire surface is straight. The more tests the inspector makes, the more reliable the test. Your own good judgement comes into use as to how many element tests are needed.

There is no assignment for this chapter. However, it is critical that you be able to describe in your own words the following definitions.

Definitions

A Geometric Control—defines a perfect model of the desired geometric characteristic. A tolerance zone is constructed around this model. For the part feature to be within tolerance, either surface elements or center features must fall within this zone.

One of Two Possible Entities Must Fall Inside the Control Zone.

Elements—are lines on the surface of a part feature. You could draw an element on a part using a pencil. It is a line running in the direction of the control.

Center Features—either feature centerlines (such as holes) or centerplanes (tabs or slots) must fall within the zone.

CHAPTER 6

Controls of Form and Profile

Controls of Form
Straightness, Flatness, Roundness, and Cylindricity

Controls of Profile
Profile of a Line and Profile of a Surface

This chapter deals with the controls that regulate the surface shape of a part: Form and Profile. They are very similar to one another as you will see.

Controls of Form and Profile Seek To Confine Surface Elements Within A Tolerance Zone

The four geometric controls of form—straightness, flatness, roundness, and cylindricity—and the two controls of profile are grouped together because they are the same kind of control. They control the shape of the surface of a feature by defining a tolerance zone in which surface elements must fit. This tolerance zone is built around a perfect model of the designed shape.

Special Case—Straightness of Centerlines or Planes

There is one exception whereby a control of form is not a surface control. *Straightness* of a feature centerline or centerplane may be controlled using form. This center control applies to round features such as a pin, and to flat objects such as a slot. Inspection of this kind of straightness is a mathematical challenge. We will discuss methods later in this unit. All other controls of form strictly deal with surface shape.

Two- and Three-dimensional Controls

The six controls of form and profile also divide into two- and three-dimensional versions. If the control is two dimensional, it is a single/individual element control; that is, each element must pass an inspection. If it is three dimensional, the control pertains to all elements on the surface.

For example, straightness is a single element requirement. The control is two dimensional; the tolerance zone is a pair of straight lines. Each individual element must be straight within the tolerance but need not compare to any other element. In contrast, flatness controls all elements on a surface. The tolerance zone is a pair of three-dimensional flat planes into which all elements of the surface must fit.

The special case for centerline straightness of a part feature that is round yields a tolerance zone that is a cylinder, or sometimes referred to as a diametral zone.

All Controls Of Surface Form Are Floating Template Controls—Concept #4

A similarity between all four controls of form is the concept of a "Floating Template" control—the control conforms to the surface as it is found. To fully understand this idea, consider what you already know about straightness. The concept we are about to learn will apply to all form controls.

How would you inspect the part in Figure 6-1 for straightness? If you answered "place the part on a layout table then run an indicator along the part in the direction of the control element," you must understand what this test would verify and the limitations. Although useful to some extent, this test has some

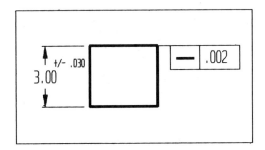

Figure 6-1 How would you inspect this part for straightness?

serious problems. Straightness is a control without any relationship to the outside world. A part is not straight as compared to a datum or a layout table. Straight is simply straight; all by itself.

If, after sweeping the indicator along the part surface, the Full Indicator Movement (FIM) did not exceed the specified total for straightness (in this case .002), you could say with confidence that the part was within the tolerance but you could not determine the actual straightness of the part from this test. Suppose the detected FIM was .003 inch. This might simply be due to a tilt of the part with respect to the table—not a problem with straightness.

An improved method to inspect for straightness would be to place a perfect straight edge template against the surface in question and check for gaps from the perfect template. If there are no gaps beyond the tolerance, the part is acceptable at that test element. In the case of Figure 6-1, the gaps must be less than or equal to .002 inch. As a refinement, the gaps could be determined using an indicator probe in relationship to the "floating" straight edge template.

This straight edge test might find that a part originally rejected by the indicator and layout table method is actually acceptable. In fact the original layout table test might have actually been showing that the part was tilted and not out of straight at all. We will see in Unit IV that even this test has a subtle limitation and must still be refined for close tolerances but it is a better method than the indicator and table test.

The difference in the two methods should illustrate the concept of floating template evaluation of surface elements. The better test of the two above was when we aligned the template to the surface in question. By ANSI definition, straightness (and all form controls) is a comparison of the feature to a perfect counterpart of itself.

Further, consider Figure 6-2. Based upon the print in Figure 6-1, and the height dimensions shown, is this part acceptable for size? The answer is yes. Now suppose you place it on the layout table and test for straightness with an indica-

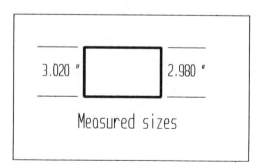

Figure 6-2 This part is perfectly straight on top.

tor. For argument's sake, this part is "perfectly straight," yet the result would still yield a full indicator movement FIM of .040 inch because the part is tilted—not out of straightness tolerance.

A floating template requires no datum nor outside influence. It simply conforms to the controlled surface wherever it is found. All controls of form then are a test of surface elements and the surface in question becomes its own reference. With controls of form, we compare the actual produced surface to a perfect counterpart of itself.

Of all 13 controls, *form is the only feature control group that requires no datum reference*. To test the part in Figure 6-2, we would use the same ground straight edge to test for gaps against the surface in question. There would be none because the part is truly straight and we have made a correct test of the surface.

Computers Test Form Mathematically

A coordinate measuring machine (CMM) (Figure 6-3) can test for straightness by running an electromechanical probe along an element. Even though this probe is connected to a rail that appears to be an outside reference—a datum—there is a big difference. The difference is that the CMM derives the true element from the data points collected. The CMM is indifferent as to the angle at which the line lays; it computes the average line then seeks the deviation. The CMM is using the concept of a floating template but it determines the answer mathematically.

The CMM is the best practical method available. While beyond the scope of this introductory chapter, there is a further, subtle reason why the CMM is the best practical method of determining the straightness of an element. Why is this type of evaluation better than the straight edge test? In Unit IV—Chapter 17, Geometric Theory, we will see the reason why.

Straightness—Concept #5

There is more to be said about straightness. Remember that straightness defines a tolerance zone—a pair of theoretically straight lines. Individual elements may deviate in any way as long as they do not violate this tolerance zone. The tolerance zone fits around the element—a floating template.

Straightness is a control of single elements. Consider Figure 6-4, which is based upon the original part that is 3.00 +/-.030 inches tall. In this example, the top of this part is straight but the surface is not flat. Each element would not compare to any other, yet, since each element would pass the straight edge test, it is a straight part. Straightness is an individual element-type test.

Based upon design requirements, the straightness control direction is shown

46 ☐ Unit II The Working Geometric Concepts

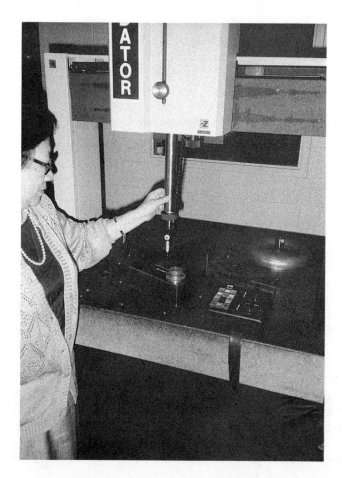

Figure 6-3 A computer coordinate measuring machine *(photo courtesy of Brown & Sharpe Corp.).*

in the view that would show the element from the side. The controlled elements are straight lines running in the direction of the control symbol.

Inspecting Straightness

The "indicator–layout table" test described above is a valid test as long as you understand what the limitations are. However, *if the FIM is within the specified control tolerance, the part is acceptable.* You have determined compliance with the tolerance but not the true value of the straightness.

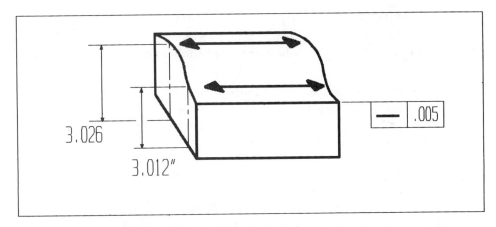

Figure 6–4 This part is straight in the control direction. It is not flat.

The "straight edge template" is an improved test for straightness. Detecting gaps away from the perfect straight edge will show the single element straightness. This test shows the actual value of the straightness.

Computer form testing methods are the best of those common to manufacturing. These methods are the best when the need for the actual value of the straightness is required or where very close tolerances are a must. These types of tests are not limited to outside contact template alignment. The computer is able to "see" the entire element being tested, then calculate the fit of the narrowest tolerance zone possible around this element.

Accuracy of computer methods depends upon many factors including surface finish of the object, probe type, and, most importantly, the density of collected data that represents the true element. Future improvements in CMM and other technical methods of form evaluation depend upon solutions to the aforementioned problems.

ANSI Rules of Application

Within the ANSI standard there are Rules of Application. They guide the user as to how to apply the geometric concepts.

There are two rules that apply to *all* controls of form.

1. No Form Specified
2. Perfect Form Envelope

Let's see how they apply to straightness.

Geometric Rule of Application Number 1

Rule 1 deals with situations where no form control has been specified, such as in the block in Figure 1-1 where the height tolerance controlled the flatness and parallelism of the top to the bottom.

Where No Form Is Specified, Limits of Size Determine the Form Control—If a print does not specify a form control then the limits of size set the extent of the form as well. Suppose on Figure 6-1, there was no callout for straightness. What would the straightness requirement then be? The answer is .060 because the surface could deviate that much without a straightness control. There is a further modification to this rule—see the following paragraphs.

Geometric Rule of Application Number 2

Another limit on straightness (and all controls of form) is based upon the concept of functional fit. Consider a pin inserted into a perfect model hole (Figure 6-5). The pin can vary but for this example the hole is a perfect cylinder 1.00 inch in diameter.

Perfect Form Envelope at Maximum Material Condition (MMC)—(For now, Maximum Material Condition means simply the pin with the *most* material allowable under the size and tolerance on the drawing. In terms of functional fit, MMC is the tightest-worst case for assembly. We will encounter this term again and study MMC in greater detail in Chapter 9.)

The pin in this example is required to have a minimum clearance to the hole of .0002 inch per the design function (.0001-inch clearance per side). That means

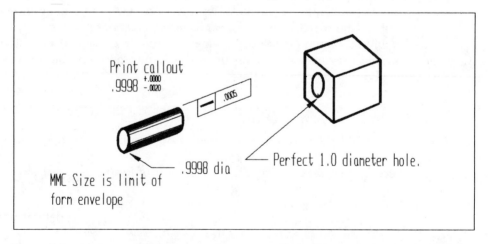

Figure 6-5 As size reaches MMC size, form is limited to perfect functional envelope.

that the functional hole would be .9998. The pin must be .9998 or smaller to fit the design criteria.

That then forms the perfect design envelope—the largest pin condition possible is MMC and no element may protrude out beyond this perfect envelope. Can you see why?

A pin at .9998 inch (MMC) could not then have any element protrude out beyond this envelope. The .9998 diameter is the true design size and shape. Elements could bend inward but not out.

The straightness tolerance is still available at MMC—.9998 diameter. But only to the extent that the elements must go inward, away from the tolerance zone boundary. Therefore, when form is being controlled, as the size dimension approaches MMC, any form deviation is stopped at the un-violatable boundary. As the size departs from MMC (in this case gets smaller), the form tolerance does not change but elements may vary outward as long as they do not go beyond the perfect form envelope. Functional gages aid this type of inspection.

Perfect Form At MMC Not Required—Since this type of form control can become complex and costly to verify, when the designer determines that perfect MMC form is not necessary, a note may release the shop from this rule (ANSI Y14.5–M 6.4.1.2.).

Special Case—A second method of releasing the form from a perfect envelope is to specify the straightness control on the centerline, not the outside surface. For more information on this special case, see ANSI Y14.5–M 5-6.4.

Computer Assisted Form Envelope Evaluation—Although perfect form at MMC is simple in definition, often in practice (for even a simple part such as this pin example) we must have some type of functional gage or advanced inspection equipment to analyze these envelope violation situations. Evaluating form envelopes is very difficult using standard inspection equipment.

A computerized profile-form testing machine or a special optical device can help determine the perfect envelope and then detect violations of this perfect form envelope. Figure 6-6 shows such a form testing machine. There are many varieties of these evaluation machines. Those that touch the part with a probe are tactile methods and those that use optics to verify form use non-tactile testing.

Straightness Review

- Straightness is a single element type control.
- Straightness is a floating template control, requiring no datums nor outside influence on the inspection.
- Straightness can be inspected using an indicator sweep test but this pro-

50 ☐ Unit II The Working Geometric Concepts

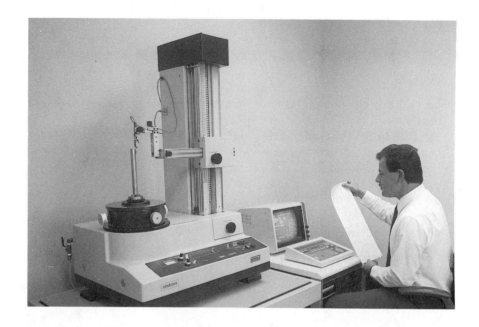

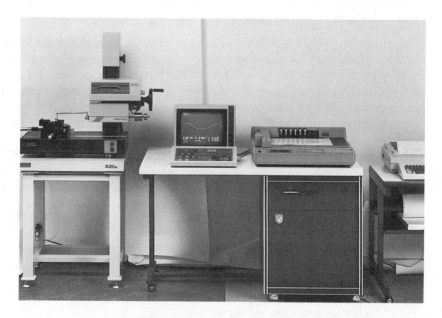

Figure 6–6 Form and profile testing machine (*photos courtesy of Brown & Sharpe Corp.*).

cedure has limited value. It can tell you if the product is within tolerance but not the value of straightness.
- If there is no control for straightness, then the tolerance limits of size limit the straightness.
- Elements may not violate the perfect form envelope unless special permission is given on the design.

We will build upon these ideas in flatness and all other form and profile controls.

Special Rules that Apply to Straightness

Abrupt Change Control—There is a special case where we may wish to further control the amount of change in any element per unit length—for example, no more transition than .01mm in 25mm length (.01/25). See ANSI Y14.5-M 6.4.1.4.

Limited Surface Control—The designer may not wish the control to apply to the entire surface. This is possible, too, with a confining note on the drawing.

Straightness Can Be Bi-directional—The designer may not wish to go to the expense of controlling flatness but still may want straightness controlled in both directions of a surface. This is possible, less expensive than flatness, and can be given different tolerances in each direction.

An example of a bi-directional straightness requirement would be in designing a bed knife for a paper shear. It must be straight *along* the cutting edge and less straight perpendicular to the edge but still must be controlled. This is a good illustration of the geometric advantage. This bi-directional straightness control fits the design requirements and is less difficult to machine. Flatness would be the next higher level of control.

> *Remember: Figure 6–7 illustrates a straightness control as applied to a centerline. Note that the control is placed adjacent to size information not as an extension of the surface. This signifies that the control applies to the centerline not the surface.*

Flatness—Concept #6 ⌗

Flatness is a three-dimensional version of straightness. Flatness is an ALL element test. The part shown in Figure 6–4 would fail a flatness test because many of the elements would lie outside the control tolerance.

The tolerance zone for flatness, then, is defined as a pair of perfectly flat

52 ☐ Unit II The Working Geometric Concepts

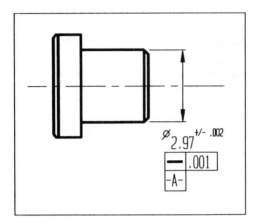

Figure 6–7 A straightness control on the center line of a part feature.

surfaces built around a perfect counterpart of the surface in question (Figure 6–8). ALL elements upon the controlled surface must fit within this control "sandwich." The distance between these control surfaces is the flatness tolerance shown in the control frame.

As with all controls of form, flatness is also a floating template control.

Inspecting Flatness

While the *concept* of flatness is easy to understand, inspecting for flatness tolerance compliance can be difficult and determining the exact value of the flatness on a real machined item even more complex. Once you understand how to verify how flat a part is, you will have a much firmer grasp of many geometric form and contour concepts.

Figure 6–8 is a geometric print controlling flatness. All elements on the top surface must fall within the control zone which is .010 inch across. Typically, as in Figure 6–9, to inspect this part, it would be set on an inspection table and an indicator would be swept across the entire controlled surface. You would accept this part as being flat within tolerance as long as the full indicator reading (FIM) does not exceed the given .010 inch.

Flatness Trap

This is a common, valid test as long as the FIM is within the .010 flatness tolerance. From this test you can say that the part is acceptable. It is within compliance. But you must understand the following: *this process tells you if the part is*

CHAPTER 6 Controls of Form and Profile ☐ 53

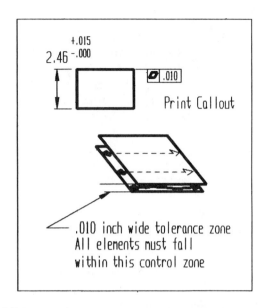

Figure 6–8 Control of flatness.

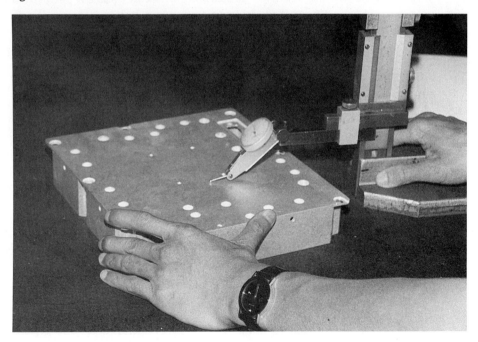

Figure 6–9 Testing flatness.

within tolerance for flatness but not what the true flatness actually is. This is similar to the straightness trap discussed previously.

Now suppose that you have produced a perfectly flat and parallel part. Imagine that you put a .012" shim under one corner of the bottom, resting on the inspection plate, and perform the same test by sweeping the indicator over the surface. The indicator would yield an FIM of .012" from the lowest element to the highest, yet would it not still be flat on top? Your answer is "yes."

Actually, since you have a .015-inch tolerance for height, this part might still be perfectly acceptable (within both flatness and height tolerance) but you need to perform a better inspection of the flatness. Flatness has nothing to do with the parallelism to the bottom of the part. No datum is required to verify flatness.

How Do You Determine Flatness?

Flatness is similar to straightness in that it is a "floating template" control. It was easy to see how to inspect straightness. You simply confirmed that the part conformed to the template within tolerance.

Ideally we do the same for flatness, but you can't place a perfectly flat plate on a part and see under it to determine gaps, can you? However, if you did do this, would it not conform to the surface—a floating template? Any inspection method used must conform to the surface then detect any gaps from the average surface.

Flatness Inspection Procedures

Indicator on Height Gage—The indicator/inspection table test is of limited value but it is often used due to its simplicity and convenience. If the resulting FIM is less than the control tolerance, the part is acceptable. However, as we have just seen, any FIM greater than the tolerance does not necessarily determine that the part is beyond tolerance. The part may be tilted and also out of flat, but once the tilt is removed, the FIM may be found to be acceptable.

More testing may be needed if the surface fails a table/indicator test. To refine the table/indicator process we could then use the "leveling method"; that is, bring three points on the surface up to some level condition with respect to the table. Then test for the FIM across the surface. This is a valid test but time-consuming and still not a perfect method. Again, it will tell you if the surface is in tolerance but not the value of the variation. You may still have some tilt in the averaged true surface (nominal plane—see below).

Why is the level method not a perfect test? Envision that there exists a perfect theoretical sandwich tolerance zone that has closed down to minimum thickness surrounding the entire surface being tested. Would not these parallel imaginary planes assume a given angle in space with respect to the inspection table, as established by the surface irregularities of the part? We will call this the nominal plane attitude. This nominal plane would therefore be established by the part feature irregularities.

The leveling process only works if you happen to select three leveling points that are exactly parallel with the nominal plane. If not, the nominal plane is still tilted to some extent and that tilt would cause some false amounts of FIM. In practice, this tilted nominal plane effect might be negligible for surfaces that are close to flat and with tolerances not too restrictive. Therefore, the leveling process can be used as long as you understand that it is an imperfect method. If a very closely toleranced surface fails a leveled flatness test, the part might still be flat. The nominal plane tilt may be contributing some error.

Another process is the "inverted method," where we lay the part feature to be tested on a flat surface such as a ground plate. Now the part is conforming or floating to a template. An indicator probe is protruding through a hole in the surface and is in contact with the part surface. Sliding the part around would show the FIM of the irregularities as compared to the surface upon which it lies. This test has a subtle limitation that we will study in Unit IV. However, this is the most reliable of the standard methods. Computers are a better solution.

Computer Coordinate Measuring Machine—The CMM deals with flatness by taking sample touches on the surface in whatever position it is found. The more touches taken on the surface, the more accurate the test will be. After all the touches are taken, the computer mathematically averages them out and fits them into the best possible flat plane. The deviation is noted and the real flatness is found, assuming that you touched enough places on the part to represent the entire surface. The CMM works from the nominal plane, detecting the thinnest possible sandwich that will contain the surface data. This assumes that enough data is provided by the user to represent the entire surface and its associated irregularities. This entire process requires many calculations made by the CMM.

Optical Methods of Testing Flatness

There are three light/optical methods for finding flatness.

Optical Flats—One very clever method of testing flatness is a specially ground optical prism. Setting the optical flat on a clean machined surface will refract (bend) the reflected light in such a way that lines will appear to the user on the flat. These lines will represent the flatness of the surface from which the light was reflected. They will be straight if the part is flat and show deviation if it is not. Comparing the optical lines to a chart, the user can determine how much deviation from perfect exists.

This method has two limitations: 1) it is limited to the total area of the optical flat, and 2) the optical tool is resting on the outer points of contact with the part, which may not be parallel to the average plane. It can't be used for large surfaces and the parts must be clean and reflect light.

Optical Laser Methods—There are methods similar to surveying that can evaluate flatness. These methods depend upon either an optical target or a laser target sitting upon the surface to be tested. These methods depend upon the fact that light travels in a straight line. If we swing a laser or viewing scope around a precision bearing, it will describe a flat surface. It then remains to align or test the surface against this flat standard. These methods are common in the aircraft industry when setting up or checking large assembly tooling. The laser is used to verify the datum frame references from which the tooling is built. Optical/laser methods are used for large areas that must be flat but they are not practical for small part inspection.

Lasers are also used in a process called interferometry. A beam of laser light is divided into two parts. One part is then reflected off a flat standard and the other part off the object to be evaluated. The returning light is recombined and the resultant interference patterns are analyzed. This costly method is not common in shops at this time but is used in CNC (computer numerical control) machine manufacturing and setup.

Flatness Review

- Flatness controls ALL elements within the tolerance zone surfaces. The distance between these tolerance zones is the flatness tolerance.
- Flatness requires no datum for verification. It is a control that conforms to the produced surface. In this book this is defined as a "Floating Template" control.
- Because of the "Floating Template" concept, indicators on machine slides or surface plates have limitations when inspecting flatness. However, they can be used if one understands the limitations.
- Where there is no flatness callout, flatness is controlled by the limits of size. Perfect form at MMC is also in effect unless released by a note.
- Flatness also controls straightness for no element running in any direction may vary beyond the control tolerance.
- Flatness is more costly to machine and to inspect. For this reason, student engineers are cautioned to analyze the function of their designs and use straightness instead of flatness if at all possible.

Challenge Problem 6-1

Answers follow.
   Refer to appendix drawing—Stabilizer Bracket
   1. How many controls of form are on this drawing?
   2. Which features must be round within .003 inch?
   3. Which datum feature has a form control on it?
   4. Which form control is also a multiple control?
   5. Which feature must be flat?

Answers to Challenge Problem 6-1

   1. Three (two controls of roundness and a single flatness control).
   2. The .500 diameter boss and the .312 diameter hole.
   3. Datum -B- must be round within .003 inch.
   4. The flatness of the face of the wing must also be parallel to datum -C-.
   5. The inner surface of the left wing.

Roundness—Concept #7 ○

Roundness is the same as straightness with the obvious difference being that the perfect model is a circle and the tolerance zone a pair of perfect circles, not straight lines. In Figure 6-10, the distance between the control circles is the roundness tolerance. If you could take a straightness control and wrap it into a circle, you would have a roundness control.

Similar to straightness, then, roundness is an individual element with two-dimension control. Consider the tapered bearing roller shown in Figure 6-11. Is it round? Yes. Each element must be round if it is to work correctly yet each element would not compare to another, would it? Each element is its own test.

Inspecting for Roundness

Similar to straightness, roundness is a floating template control. A radius gage is a floating roundness testing tool. It conforms to the outside of the curve and then gaps are detected. But since it only covers 90 degrees of the curve, it is not very useful for roundness all the way around the part. The best and simplest test of roundness is a ring gage. The drawback for ring gages is that one must have many sizes to be of general use.

The roundness tests below are common but each has a limitation. With a geometric understanding each is useful. Again we find that it is relatively easy to

58 ☐ Unit II The Working Geometric Concepts

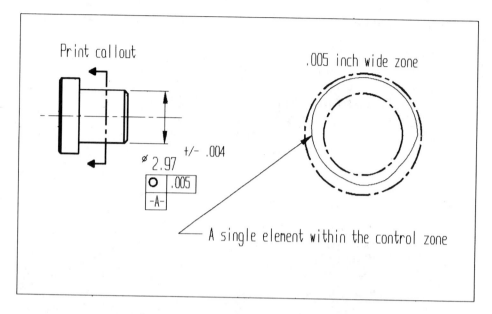

Figure 6–10 Control of roundness.

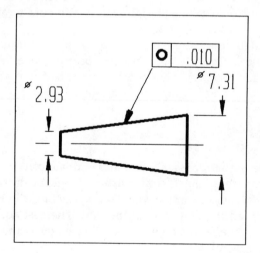

Figure 6–11 This tapered pin must be round.

test for compliance—is the part within tolerance? But it is far more difficult to arrive at an exact value of deviation.

Micrometer Test—When you use a micrometer to test a round part, what are you actually determining? If you answered "the thickness across" you are correct. The distance across the part is not the roundness. In Figure 6–12, note that distance A and B are equal. A micrometer is by far the most common method used to test for roundness, but remember it can be fooled.

Tri-anvil Micrometer—This micrometer has two bases in a triangle with the spindle for three-point suspension. It would find the out-of-roundness part in Figure 6–12 but there are other shapes that could fool these tools too. This is better than a plain micrometer for finding roundness.

Using both a standard and a tri-anvil micrometer on a single roundness element would detect any problem.

Vee-block and Indicator—By now your geometric sense should tell you what is questionable with this process. Laying a part in a vee block and rotating it against an indicator is similar to the tri-anvil micrometer and has exactly the same limitation. There are classic shapes with lobes and flat spots; they are obviously not round but when placed in vee blocks (or tri-anvil micrometers) and rotated against, an indicator will show zero error! The most common of these false round shapes is called a tricoidal (see Figure 6–12).

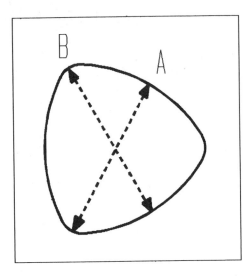

Figure 6–12 Distance A = Distance B, but this shape is not round! It is a tricoid.

60 □ Unit II The Working Geometric Concepts

Rotating the Part Between Centers with an Indicator (in a Lathe or Fixture)— This test has the same limitations as when testing straightness by moving the indicator in a straight line on a table—you are introducing an outside influence on the test. Roundness is a floating control with no need for datums or outside comparison—in effect, comparing the surface elements to a datum, the center of the rotating chuck or fixture.

In this case, the datum is the axis about which you are spinning the part. You are now moving the part in a circle. Consider Figure 6–13. Diameter B is perfectly round and is connected to Diameter A. We hold the part true on Diameter A and spin the part—what do we see? A large FIM, yet the surface being tested is actually round. Again, this test then introduces an implied datum—the center of rotation. You are actually testing the radial distance from the axis with this method which is not roundness! We will soon learn this to be runout—comparing the surface to a datum axis, the center of Surface A.

As with previous indicator-FIM tests, if this test shows no more FIM than the control tolerance, it means the part is acceptable. It complies with the roundness control, but does not tell you what the roundness actually is. What you may have found is that the test diameter is actually eccentric (off center) from the axis of

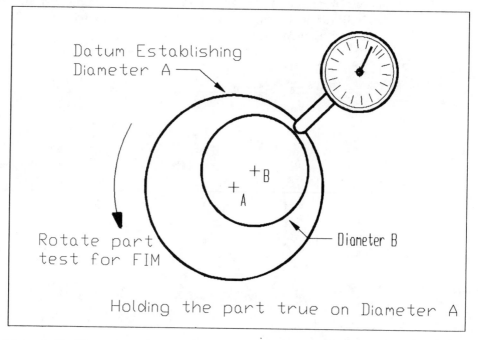

Figure 6–13 This test can determine if Diameter B is within the roundness tolerance. But it *cannot* detect the value of the roundness. Diameter B is round but not on center of Diameter A.

rotation. You might, with practice, be able to discern the difference by watching the progression of the indicator needle.

Computer Measuring Machine—The roundness measuring machine is the most accurate method. There are various levels of roundness testing equipment. Some simply reproduce an expanded template or printout of the element tested while others analyze elements and report the minimum pair of theoretical perfect rings that can contain a tested element. The distance apart of the calculated rings is the actual value of the roundness. In these roundness testers, the computer mathematically applies the same floating template concept as with flatness and straightness. It uses the data to create the nominal circle and test for deviations.

You can see that except for a ring gage and computer/roundness testing equipment, all the common roundness test methods used were not quite correct but can be used if you understand what the results are actually telling you. If an element fails, it may mean that it requires more testing. Computer inspection equipment is a practical way of determining the actual roundness value.

Geometric Sense Required—In shop practice, you must consider several facts in a standard roundness determination. Consider the nature of the feature (length and relative diameter); the process and the equipment upon which the round object was produced also can be considered as part of the test. With some experience many objects can be measured with a micrometer and the result be trusted. Close tolerances and objects that might chatter or lobe during machining are then subjected to stricter tests for roundness. The machine history and the machinists are also considerations. It is proper to use your own best judgement based upon your geometric understanding of the manufacturing conditions and of the test limitations.

Most of the roundness tests above have some type of limitation; however, as long as you understand these limitations, the tests are valid and useful today. The entire subject of geometrics will thus continue to challenge your thinking. The purpose of this discussion was primarily to get you thinking geometrically and to understand form controls.

Form seems simple yet it can be difficult to inspect. The more you understand the subtle differences and limitations in the testing procedures, the more control you will have.

Roundness Review

- A roundness control is defined as a pair of perfect circles into which an individual element must fit. The distance between the circles is the roundness tolerance.
- Roundness is an individual element control. Each element is its own test and need not compare with any others. It is a two-dimensional test.

- Roundness is a floating template control. Roundness tests need not have a datum nor any machine/layout table to test it. It requires no outside influence. Round is simply round to itself.
- Where no geometric roundness control is called out on the print, the limits of size will determine the roundness.
- Roundness is a test of surface elements only.
- Perfect form at MMC applies unless otherwise released.

Cylindricity—Concept #8

Cylindricity is the three-dimensional extension of roundness. The relationship between straightness and flatness compares to the relationship between roundness and cylindricity. Cylindricity is an ALL element control. If you could take a flatness control and wrap it into a cylinder, you would have a cylindricity control. Cylindricity also controls roundness and straightness and also the parallelism of the straight elements. These extra controls within an individual control are called "Embodied Controls." As we continue our studies, you will see many embodied and related controls within the system.

The cylindricity tolerance zone is a pair of concentric cylinders in which ALL elements of the controlled surface must fit (Figure 6–14). The distance between the tolerance cylinders is then the cylindricity tolerance (space on one side).

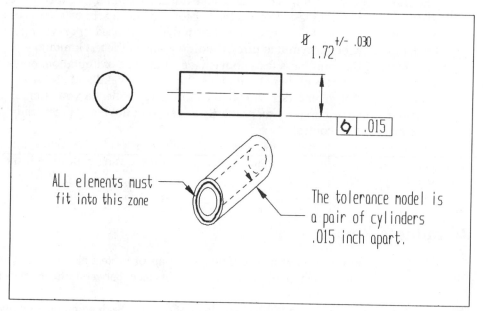

Figure 6–14 A control of cylindricity.

True of all controls of form, cylindricity is a floating template control. Also if there is no cylindrical control specified, the tolerance for diameter limits the extent of cylindricity.

Inspecting Cylindricity

This is the most difficult of the form shapes to inspect. It is common to spin the part and use an indicator—which, as you already realize from the roundness development, only tells if the part is within the cylindrical tolerance but not what the cylindricity actually is. If the rotating test fails, it does not mean the part is necessarily out of cylindrical tolerance.

It may mean that you are spinning it about an axis that is not the true center of the form you are inspecting. Ideally, we could slip a cylindrical ring gage over the part then test for gaps but that is not possible due to the closed shape—you couldn't see under the gage. Even if the gage was made of some clear material, to see the gaps, you couldn't measure them and you would need one size for every possibility.

Micrometers and all the methods discussed under roundness have the same limitations when testing for cylindricity. It is most common to accept that if various elements are *round*, are the same size from element to element, and the elements running parallel to the axis are straight, then the part is cylindrical—but now you understand that that may not quite be true.

In practice, the process and equipment with which the part is machined is often considered when evaluating cylindricity. If the part is turned and/or ground on quality machines, it is accepted that if it can be tested to be round and straight on the sides, with no taper, then it is cylindrical. But keep in mind the limitations of finding roundness in the first place.

Again, a computer profile/form test is a nearly error-free way to verify cylindricity. Here the part is spun against a probe or vision system and the data collected for a number of elements along the surface. The computer then finds the "best fit" cylinder and then computes the deviations away from the cylinder. The computer compares the surface of the part to a floating cylindrical template and detects the deviation from that best-fit model cylinder.

Cylindricity Review

— Cylindricity is an ALL element control.
— Cylindricity is a floating template control.
— Cylindricity is defined as a pair of theoretically perfect concentric cylinders into which the entire element set must fit (the entire part surface).
— Cylindricity is more expensive to produce and inspect than roundness.
— Without a cylindrical control, the limits of size take over.

- Perfect form is required at MMC whether it is a pin or a precision hole unless released by an engineering note.
- Cylindricity is a surface element control only.
- Computer evaluation is the best method of inspection; however, standard methods are acceptable as long as the user understands their limitations. The most common tests using indicators, micrometers, and straight edges can tell the user that the part is within cylindrical tolerance, but not the value.

Controls of Profile

Profile is Similar to Form

Within the ANSI Y14.5-M geometric system, there are two controls of Profile. They are Profile of a Line and Profile of a Surface. Both are controls of surface shape of features. Profile of a line is a SINGLE element (two-dimensional) control and profile of a surface is an ALL element (three-dimensional) control. Instead of straight or round, the shape is now defined.

Profile, as with all geometric controls, defines a perfect model around which a tolerance zone is defined. Elements then must be contained within the zone for the part to be within tolerance. By bending a straightness control into some shape, you would have a profile of a line control. There are two restrictions on the bending. First, you must keep the entire control zone in a flat plane. That is, the individual bent element could be drawn on a flat piece of paper because it is a two-dimensional control. Second, in geometric control of profile, you must be able to define the shape.

For surface profile the bends are as though you bent a sheet of metal—they only bend one way. This is called a simple shape. There will be more on this subject later. Can you see the similarity between straightness, roundness, and profile of a line? How about flatness, cylindricity, and profile of a surface?

Profile Is Different from Controls of Form

There are two reasons why profile is separated out from controls of form.

Difference 1.—The model is a shape that must be defined in some way. When dealing with the controls of form, you did not need a definition of straight or round, did you? Profile, however, deals with shapes that require some kind of definition.

Difference 2.—Profile may be a floating template control or it will more likely be a fixed template (referring to a datum) control. Depending upon the function of

CHAPTER 6 Controls of Form and Profile □ 65

the profile, the shape usually needs to be kept in some orientation to other part features. The orientation of the profile requires datum reference.

Floating Template Example

In Figure 6–15, the shape definition is given without reference to a datum. Therefore, the floating template concept may be used to inspect the part. The designer simply wants the part to have this profile but does not need it to relate to any other part feature. This is similar to straightness and roundness. The inspection template for the shape conforms to the surface being tested.

Floating profiles are less common but when dimensioned without datum reference, there are only three practical ways to inspect them.

1. Create a template that has the correct outside shape then use it to test elements. This is the same as a straight edge for straightness or a radius gage for roundness. These checking templates are usually machined on CNC equipment but may also be made by standard layout and cut methods.
2. A CMM computer is programmed with the description of a profile element. It can mathematically compare each element against this standard. The CMM floats the description to the elements as found. This requires advanced CMM capability. The expense and time required for this ad-

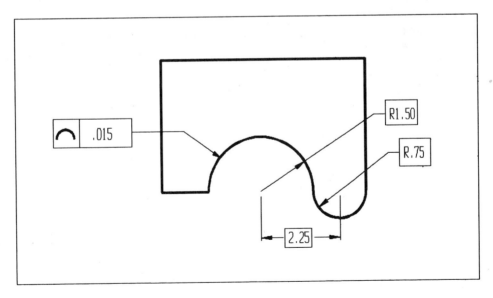

Figure 6–15 A non-datum related profile.

vanced method could be justified on precision profiles such as aerospace parts where designs are complex and tolerances very small.
3. Use a height gage and indicator on an inspection table. We will discuss inspection of profile more later. For now, this time-consuming method works as long as the user can move the high gage accurately across the part and know what the vertical height ought to be at each stop along the profile.

Fixed Template Example

If the shape is defined from datum reference points, then the inspection is done using a fixed template (Figure 6–16). Here, the template or inspection method must relate to an outside influence. In the example, the axis of the shape must be parallel to datum -A- within .005 inch.

Elements in Profile Controls

Profile of a line is a two-dimensional—SINGLE—element control and profile of a surface is three-dimensional—ALL elements. This relationship is the same as that between straightness and flatness or roundness and cylindricity. Profile is still very similar to form except we are now dealing with some shape that must be defined and it most likely will need to be related to a datum.

Within either profile of a line or of a surface, the elements will be the same shape along the entire controlled feature. They can vary only to the extent of the control tolerance.

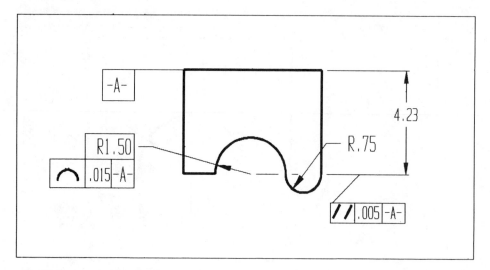

Figure 6–16 A profile related to a datum.

Establishing the Profile Element

The shape along each element may be quite complex. It could be composed of circles, straight lines, or other mathematical curves such as helixes, parabolas, or ellipses, or the curve could have no mathematical counterpart—a free shape. This free shape line must be described as a set of coordinate points along the curve or be provided as a template/model. However the profile shape is established, it will remain a constant across the entire surface that is being controlled. Again, the single element must be restricted to lie wholly upon a plane, as though it were drawn on a flat piece of glass. It can be shaped in many ways but must be on a flat plane in one direction.

Methods for Defining Shapes

The perfect template for a single element may be defined in four ways:

1. In standard drawing terms—radii, lines, and points. With or without datum reference.
2. In coordinate values—also called a table of offsets. In the aerospace industry these sets of coordinates that define a profile shape are called master dimension identifiers (MDI).
3. By mathematical formula. Or sets of formulae.
4. By templates. Exact functional shape samples.

These are either Solid Master Models or Drawn 2-D Templates.

A drawn template is often referred to as a PCM. PCM means Photo Contact Master. It is a drawing on mylar plastic film. Mylar has good temperature and moisture stability. It is a perfect reproduction of the designer's original shape determination.

A master model is a developed shape template or a solid model. It establishes the profile element and can become the inspection template. The part shape is compared to this template.

Profile of a Line—Concept #9 ⌒

Profile of a line is a two-dimensional, single element control. The tolerance zone is a pair of lines defined around the perfect shape (Figure 6-17). The distance between these tolerance lines is the profile tolerance.

The inspection test would be for each element and would not relate to any other on the part. Compared to any single element, an element elsewhere on the surface could be higher, lower, or tilted in relationship. But both elements must conform to the test template within the control tolerance. If the control was related to a datum, such as in Figure 6-16, then the axis of the element must comply with the controlling relationship to the datum. In Figure 6-16, the axis must be found to be parallel

68 □ Unit II The Working Geometric Concepts

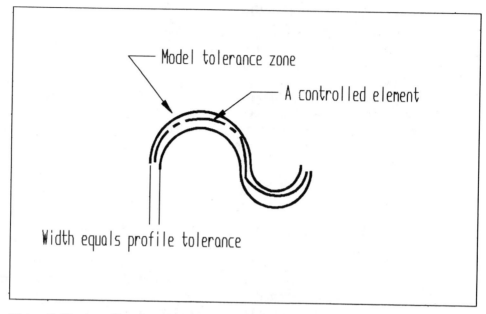

Figure 6–17 A profile tolerance zone.

within .005 inch to datum -A-. If there existed a height tolerance to the 5.16 dimension, then the axis could also vary in height above datum -A-.

Outside Mold Line–Inside Mold Line Application

Profile of a line or surface may be applied bilaterally; that is, the tolerance zone will be equally spaced around the perfect form or it may be unilaterally applied.

Unilateral profile is either outside or inside the given perfect profile. This is often used in mold making where an inside (IML) and outside (OML) part must nest together and the gap be controlled to a minimum tolerance. Complexities arise on OML/IML work because of the expansion or contraction of the shape as we go away from the perfect profile.

A simple example would be Figure 6–16. Suppose we were setting a .030-inch profile control using IML. The outer template would be exactly as the print describes, but the inside template would require some changes. The extent of the control zone would go inward from the perfect profile. The 1.5 inside radius would get smaller while the outside radius would become larger. The 1.5 radius would limit at .030" = 1.470" while the .75 radius would be .78 inch. On this simple shape, the center points do not change with the IML application but you can see that center points might shift too if the profile was more complex.

Inspection of Profile of a Line

The floating template method is by far the simplest method to verify profile. A checking template is made by standard or CNC methods. Hold it against the part and test for gaps.

A difficulty arises when we deal with fixed templates whose profile relates to a datum. This is due to the required relationship of the profile to a datum reference.

If you were inspecting Figure 6–16, you would no longer have a floating template situation. You simply cannot place the template on the surface. The inspection template and datum -A- must remain parallel within .005 inch. The template must be brought to the part in relationship to the datum, then the element checked.

A fixed template simplifies inspection in one way: you can now use the height gage and indicator method on an inspection plate. You would need a set of coordinate values along the profile. Look, for example, at Figure 6–18, where Figure 6–16 has been broken down into vertical components. Every half inch has

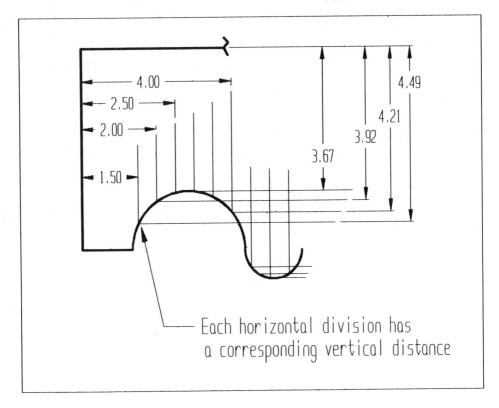

Figure 6–18 A set of profile values.

a corresponding vertical coordinate value and this could be inspected. This inspection method is common but dependent upon the ability to move the height gage laterally to some exact location.

There is one further complexity to take into account with point testing as described above, and that is the concept of normal lines. A normal line is perpendicular to a curve at a point (see Figure 6–19). When inspecting profile, you need to test normal lines from the profile because the tolerance zone projects out along this line. In other words, the distance between the perfect model and the tolerance limit is along the normal line.

When inspecting a contour, mathematical complications can arise when close tolerance IML or OML work is inspected because we tend to inspect non-normal lines. In Figure 6–20, the perfect profile is on the bottom while the tolerance zone limit is on top. Suppose this distance is .010 inch. Point A is directly above Point B. That is where we would tend to measure the error for Point B. But distance A–B is not .010 inch; it is more than .010. Distance B–C is .010 inch but slightly offset sideways from test Point B. Point C is actually the upper limit for error away from Point B. The limit for profile for Point B is at Point C. The sideways shift of tolerance limit Point B becomes slight for nearly flat curves but increases for rapidly changing profiles.

This phenomena becomes more important with tight curves and can be dis-

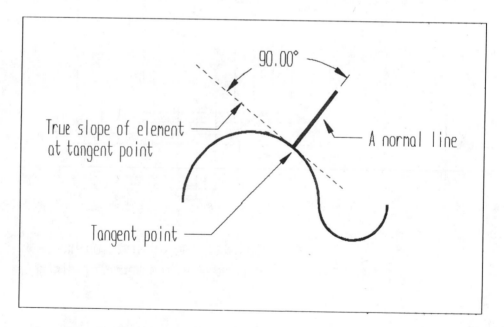

Figure 6–19 A *normal* line is 90 degrees to a curve.

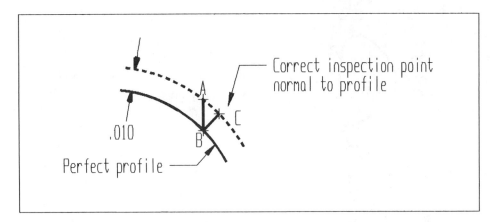

Figure 6-20 Offset inspection points.

counted on shapes that are not too extreme. But normal line shift must be considered when controlling and inspecting profile.

The most common method of dealing with normal line/point shift prior to computer evaluation methods required that the tolerance zone be decreased and the shift factor ignored. For example, in Figure 6-20, the .010-inch natural design tolerance available possibly would have been trimmed to .005 inch and the design released to the shop. This means that the manufacturing cost went up. Today, we have computer methods to set up the inspection values. This provides all the possible tolerance for the process.

Profile of a Surface—Concept #10

Following the systematic approach we have taken, you should require little in the way of explanation for profile of a surface. It is a three-dimensional, all element control. The shape is continuous along the surface—every element on the controlled surface must have the same shape and collectively they must fall inside a pair of tolerance zone surfaces.

In Figure 6-21, we see a profile of a surface being controlled on the block. All elements must lie within the control tolerance surfaces. This is a similar control to flatness where the tolerance control zone is a pair of theoretical parallel surfaces inside which all elements must lie.

With profile of a surface, the control zone is a pair of shape-defined surfaces built around the perfect shape. The shape is a constant along the surface. The shape remains the same for each test element. You can observe that there is also a straightness control within profile of a surface. Elements running perpendicular

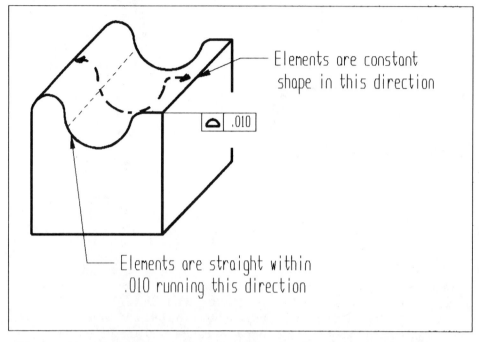

Figure 6–21 All elements must lie within .010-inch-wide tolerance zone surfaces.

to the profile must be straight within the control tolerance. That is to say that they too must fit into the shape tolerance zone.

Simple Shapes Only

Profile of a surface deals with simple shapes only (with the one exception of spherical surfaces). A simple shape is one that you could form from a piece of paper without wrinkling the paper. A cone is a simple shape—you could wrap a piece of paper around a cone. In other words, geometric shapes embody a control of straightness, perpendicular to the shape element. Complex curves are not yet addressed within the geometric standards.

Review of Controls of Profile

— Profile may be a SINGLE or ALL element control.
— The model is a shape that must be defined.

- The shape may relate to a datum or it may not. Datum related is more common.
- Inspection is floating template for non-datum controls.
- Inspection is fixed template for datum related controls.
- If the control is for profile of a surface, the surface is a simple shape. A sheet of paper could conform to it.
- Profile controls straightness in a direction perpendicular to the profile element.
- Geometric profile deals with simple shapes with the single exception of spheres. This could change with future updates.

CHAPTER 7

Controls of Orientation—Parallelism, Angularity, and Perpendicularity

Controls of orientation regulate the relationship of one feature to another. The feature used as a basis for comparison must be a datum. Controls of orientation always refer to one or more datums.

For example, if the centerline of a hole is required to be parallel to a surface of a part then the reference surface must be assigned a "Datum Identifier." To measure the parallelism of the hole centerline, the comparison surface (datum feature) must be held against an inspection quality surface (for reasons you already understand from Chapter 3). This way, the parallelism is checked against the datum surface established by the part irregularities. It would be a good idea to go back and review Chapter 3 now if you do not understand this concept.

The three orientation controls are similar. They all define a perfect model around which a tolerance zone is constructed. The zone is at a specific angle to a datum.

The Angle of Control to the Datum

The difference between each orientation control is the angle at which the perfect model and the control zone lie to the reference datum. If the control is perpendicularity, then the model is at 90 degrees to the datum. If the control is parallelism, then the model is at 180 (or zero) degrees to the reference datum. If the control is angularity, then the model is at some specific angle to the datum and must be *defined* on the drawing.

Tolerance Zone Shapes

The orientation tolerance zone may take three shapes:

1. Most common is a pair of *flat surfaces*. These theoretical tolerance zone surfaces are used to control surfaces of part features.
2. A second tolerance zone shape is a *pair of parallel lines*. For example, a tab centerline may need to be at some angle to a datum in a single plane—a two-dimensional control.
3. The third shape is a *cylinder*, if the control applies to the centerline of a hole or pin. In this case, the diameter symbol will be in the frame by the control tolerance. A zone of this shape is called a diametral zone in two dimensions and a cylindrical zone in three dimensions. (The diameter symbol is a circle with a line through it.)

Center Feature Symbols on the Drawing

When using orientation to control a centerline axis of a feature, the control frame will be drawn next to the feature dimensions and not be connected to the surface. This signifies to the user that the control pertains to the centerline of the feature, not the surface.

Single or ALL Element Controls

Controls of orientation are ALL element if they seek to regulate a surface. If the orientation control is dealing with the centerline of a hole, then it is a single line control but the line is not an element. It is a mathematically derived line that must fall inside the zone.

Occasionally there are functional design requirements whereby the entire surface does not require an ALL element orientation relationship. The designer may release the shop from achieving all elements within the orientation zone with a note such as EACH ELEMENT or RADIAL ELEMENTS ONLY (see ANSI 6.6).

Parallelism—Concept #11 //

Parallelism defines a pair of theoretically perfect planes or lines (or a cylinder if controlling a centerline) that are truly parallel to the control datum. The distance across these control zones is the geometric parallel tolerance. Surface elements or centerlines may assume any shape as long as they do not violate this zone.

A Floating Zone

Note on Figure 7-1 that the parallel zone could float up or down to accommodate the height tolerance range of .020 inch (+/- .010"). The height could vary; how-

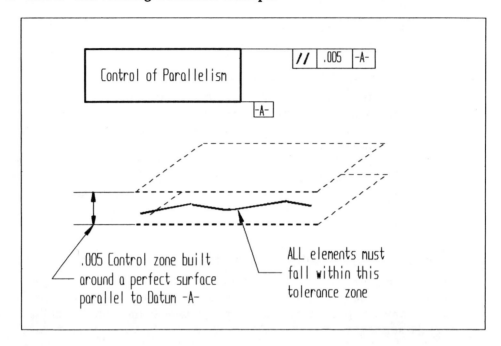

Figure 7-1 Control of parallelism. The top of this part is controlled for parallelism to datum -A-. This also controls the flatness.

ever, all elements on the surface must fit into the control zone which is .005 inch across and parallel to datum -A-.

Figure 7-1 shows a surface being controlled parallel to datum -A- within .005 inch. This then would be an ALL element control for the surface of the block. Also note in Figure 7-1 that the distance across the tolerance zone is .005 inch which is .0025 on either side of the perfect model. All surface elements must lie within this theoretical tolerance zone.

Embedded Control of Flatness

What is the flatness control for this surface in Figure 7-1? The answer is: flat within .005 inch. It is then realized that when dealing with the orientation of feature surfaces that the flatness and straightness are embedded within the control. The parallelism control limits the flatness to the same tolerance.

Could a designer specify a tighter flatness control than the orientation? Yes; it is possible to fit a flatness control inside the orientation tolerance zone. On the part in Figure 7-1, if the designer required a stricter flatness control of .001 inch, that would be possible and acceptable because, remember, flatness is a floating control. A surface with .001-inch deviations from flat could be contained within the .005-inch parallel zone.

Would reversal of the controls be acceptable? NO. It would be meaningless to have a tighter orientation control than flatness because the orientation control would supersede flatness and the tighter control would take precedence.

Inspecting Parallelism

Inspection of parallelism (or of perpendicularity) of a surface is not difficult. One must establish the reference datum against an inspection quality surface such as a layout table or ground plate then use a dial indicator to sweep across the entire surface of the feature in question. The FIM for the surface must not exceed the stated tolerance.

While inspecting a controlled surface for orientation using an indicator and FIM test, you are verifying compliance to a flatness control but not the value of flatness. The flatness is equal to or flatter than the orientation FIM result. The surface could be flatter than the orientation FIM due to tilt in the plane. As long as the plane tilts less than the orientation control, the feature is in tolerance.

Parallelism, along with the other two orientation controls, is easy to verify on a CMM. First, the user establishes a datum surface for the control. This may be accomplished by one of two methods. The difference in each method is the way in which the datum is established for the orientation.

The first uses the CMM exclusively: the operator makes enough touches on the datum feature surface to satisfy the establishment of an average surface. This requires some judgement on the part of the operator because there must be enough touches to represent the entire surface. If, for example, only a single high point is missed, the resultant datum will be incorrect. Next, a second set of "hits" will be taken on the controlled surface to represent the entire surface. The computer can then calculate the orientation of the controlled surface to the datum created by the control reference surface. This is the most efficient and simplest method of the two as long as the part features are reasonably flat, and the surfaces within reach of the CMM probe.

The second method uses the entire datum surface of the part and puts this surface in contact with a gage quality plate. This is similar to standard inspection methods using plates and indicators. Here a standard fixture, angle plate, or the CMM table is used to establish the control datum. The first step in the process would be a set of three probe touches on the contacting surface that is now the control datum. This eliminates any guesswork about how many hits to take in establishing the datum. Part irregularities are dealt with and the true datum is established—the contacting plate. The next step is to inspect the surface in question with respect to the contacting surface. While this method eliminates any guesswork about the datum, the controlled surface may still be misrepresented if the CMM operator misses the correct data points.

Orientation inspection of feature centerlines is far more complex. The major complexity is that there actually is no real centerline or centerplane. It's easy to define the centerline, but on a real part it only exists as a composite of the entire feature shape. Finding the centerline of a feature is a mathematical process.

There are solution methods that lie in using inspection pins and other tools that detect the actual geometric feature. The feature is treated as you would a datum. Some type of perfect object is brought into contact with the feature and, from this, the centerline is abstracted. An example is inspecting the orientation of a hole to a datum. The first step is to hold the part in such a way as to establish the control datum. This is done either on the CMM or using contacting plates, surfaces, or fixtures.

Next, the user needs to establish the "geometric hole." Here, a set of ground gage pins of successive sizes is used. The largest pin that just slips into the machined hole represents the geometric hole. Snugly in place, this gage pin can then be inspected from the outside because it is of gage quality—without irregularities. Testing along the pin with an indicator will provide the deviation from the orientation control tolerance. However, you must be aware of unit deviation; for example, if the hole in question is 1.0 inch deep. You must either test a total of 1.0 inch of the pin and read the FIM directly or you could test a lesser distance—say .100 inch—then multiply times 10. You could also test 2.00 inches and divide by 2. The longer test is the most accurate but be cautious not to use the total FIM, but rather find the FIM per the length of the controlled feature.

The outer surface of the pin represents the centerline axis. Can you see how this gage pin method could find the actual orientation of a crooked drilled hole? Functionally, no matter what shape and orientation the hole takes on, the largest pin simulates the tightest fit virtual condition of the hole. It is the largest perfect cylindrical space available for assembly.

Perpendicularity—Concept #12 ⊥

Similar to parallelism, perpendicularity defines a pair of theoretically perfect planes or, in the case of a round feature's centerline, a cylinder. The control tolerance is the distance across the tolerance zone. Elements or centerlines must not violate this zone.

Figure 7-2 shows a surface being controlled, thus the perfect model would be a theoretically perfect surface, perpendicular to the control datum. The tolerance control zone is then built around the perfect model. All elements must lie within the zone in order for the part to be within tolerance.

Figure 7-3 shows the centerline of a hole being controlled perpendicular to a

CHAPTER 7 Controls of Orientation ☐ 79

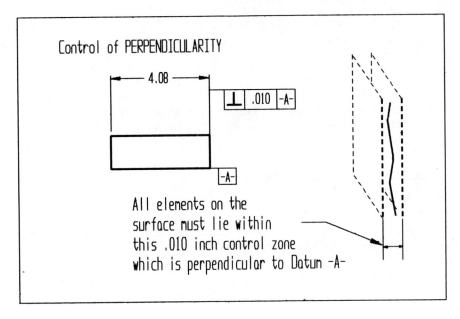

Figure 7-2 This control is similar to parallelism.

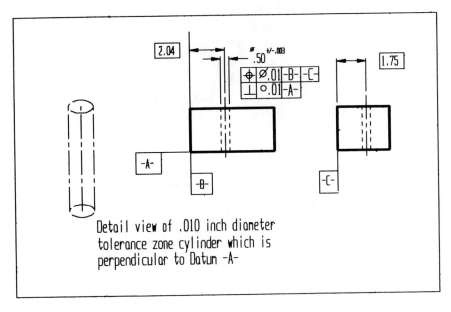

Figure 7-3 The centerline of the .50 inch hole must fall within this zone.

80 ☐ Unit II The Working Geometric Concepts

surface, thus the control tolerance would be the diameter of the cylinder. The centerline of the hole may vary in any way as long as it remains confined within the control cylinder.

Circular Tolerance Zone

Please note in Figure 7-3 that the diameter symbol (a circle with a line through it) was included in the feature tolerance box. This then shows that the shape of the tolerance zone is not the distance between parallel surfaces but is a cylinder.

Inspection of Perpendicularity

Inspection is similar to parallelism except the 90 degree relationship must be verified. Without a CMM, this then requires an inspection-quality 90 degree gage. On a layout table, the part must be held against an inspection right angle plate then tested with the indicator (Figure 7-4). Or you might check the part against an inspection-quality square.

Remember to use datum priorities here. Hold the part against the datum called out for the control then inspect the controlled surface with reference to the

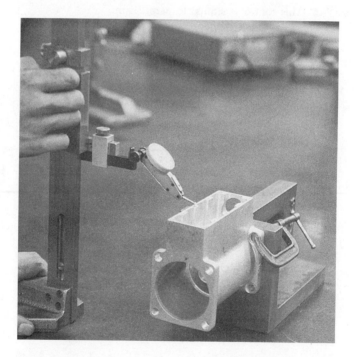

Figure 7-4 Testing perpendicularity with inspection table methods.

datum. Recall Figure 3-6, where the inspector needed to compare the controlled surface to the datum in order to get the correct result.

A CMM can inspect perpendicularity by simply comparing two surfaces (Figure 7-5). The CMM user first establishes the datum surface in any position, then evaluates the orientation of the second surface where it is found. The computer can evaluate both surfaces and the orientation between them. However, again you are cautioned that the density of data can be a real problem on parts that may be fairly irregular on either the datum feature surface or the controlled surface. If, in testing either of these part surfaces to establish the control datum, you miss the single highest point that would contact a physical datum, the result is inaccurate.

Inspecting perpendicularity of feature centerlines without using a CMM is usually accomplished with gage pins or other inspection-quality tools such as adjustable parallels for slots. The difficult part of this type of inspection is that the centerline being controlled is a composite of all the surfaces of the feature in question. Take, for example, inspecting a hole's centerline perpendicularity to a datum. The tightest possible pin that is inserted in the hole can then represent the

Figure 7-5A Testing perpendicularity using a CMM.

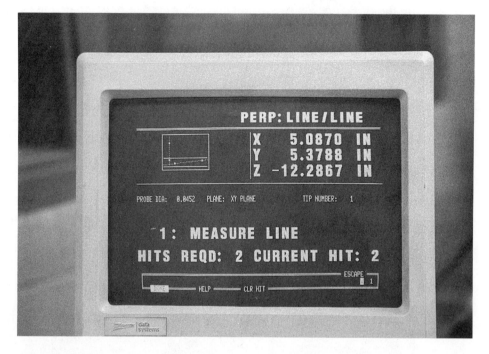

Figure 7-5B Testing perpendicularity using a CMM.

centerline of the feature. Since its surface is near perfect, the outside of the pin can be tested with an indicator.

The possible danger here is for tapered holes or features that do not yield distinct results. A CMM is one solution for this type of problem.

The CMM works inside a hole, for example, by testing several circular elements. The connected centerline of these test circular elements becomes the centerline of the hole. Data density is a matter of judgement on the part of the operator; however, more is better in terms of averaging out the true hole centerline. Feature centerline perpendicularity error is detected here where it might go unnoticed using the standard methods of inspection mentioned above.

Angularity—Concept #13 ∠

Angularity defines a pair of theoretically perfect planes/lines, or a cylinder if the control is for a centerline of a hole. The model for this control zone is at a perfect angle with respect to a datum (Figure 7-6). The distance across the zone is the

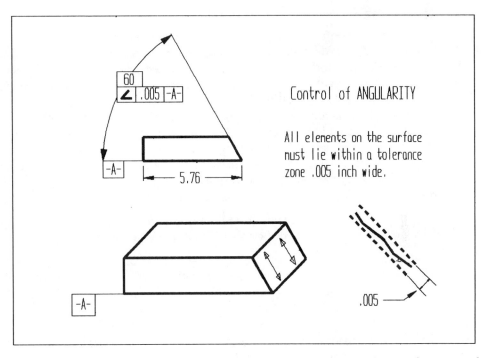

Figure 7-6 The control zone is a perfect 60 degrees to datum -A-. There is also a control of flatness embodied within a control of angularity.

geometric control tolerance. Elements may vary in any way as long as they do not penetrate outward through the zone. Angularity is very similar to parallelism and perpendicularity except in the way the angle is shown for the perfect model.

Basic Dimensions Are Used To Define the Control Angle

Note in Figure 7-6 that the angle that defines the controlled surface has a box around it. This indicates to the reader that the dimension is *"Basic."* That means it is a perfect dimension without tolerance. There must be tolerance built around this dimension but *the basic dimension has no tolerance. It is the target.*

To further understand Basic, consider this: In the case of perpendicularity when I said that the control was 90 degrees to a surface, was that with a plus or minus anything? No, of course not—you and I know exactly what perpendicular means. Parallel was also given as a basic definition. So too, then, for the angle in angularity. It is given as a perfect definition. This is Concept #18 which we will further investigate in this text.

Inspecting Controls of Angularity

Inspection of a control of angularity is not difficult. Since orientation always refers to a datum, standard layout table and indicator methods are appropriate. However, with standard methods you need to use some device that can be set to an exact angle. The more common of these are solid ground plates having commonly used angles. Another universal angle setting tool is an adjustable sine plate.

A sine plate is hinged on one edge and has a round rod at the other edge. The round rod is parallel to the hinge and the distance between the hinge and rod are a known value such as 5, 10, or 20 inches. Accuracy depends upon the fact that the length between the hinge and rod are exactly known. A calculated stack of material under the rod edge will elevate the top surface to an exact angle compared to the sine plate base (see Figure 7-9). The stack is calculated using the formula:

> Stack Height = Sine Ratio of the Angle ×
> Length Between Rod & Hinge Centers

To inspect the part shown in Figure 7-6, we need to establish a perfect angle between the inspection plate and datum feature Surface A. This should bring the controlled surface up 45 degrees—parallel with the inspection table surface. Now sweep an indicator over the surface and test for FIM.

Angle Tolerances in Geometrics Are Not Fan-Shaped

The concept of angularity may conflict with your nongeometric idea of dimensioning and tolerancing angles (Figure 7-7). You probably think of an angular tolerance as plus or minus so many degrees. This doesn't serve functional fit at all.

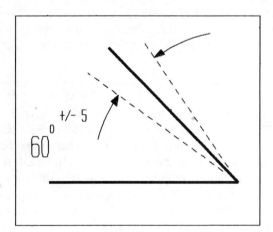

Figure 7-7 A nongeometric tolerance zone results in a fan-shaped control.

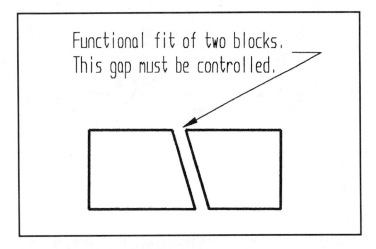

Figure 7–8 Controls of orientation serve functional fit.

As with all geometrics, angularity may be explained using the idea of function and fit. When assembling two parts that have angles, the major goal is to control the fit between them (Figure 7–8). The geometric definition of angularity does so with ease. Angularity must be similar to the other orientation controls. *There is no fan-shaped tolerance zone in real functional fit.*

In Figure 7–7, the result is a fan-shaped possibility for the shape of the part. It is very difficult to control the fit between two of these fan-toleranced parts. In fact, what most nongeometric designers usually do with angular tolerances is to tighten the plus or minus tolerance until they are confident that the parts will assemble. As you learned in Unit II, this drives the cost up exponentially and causes many production problems. (This explanation also relates to perpendicularity and parallelism.)

Template and Optical Methods of Testing Orientation

There are means other than the inspection table and CMM for testing orientation. The most common perpendicularity test is a fixed master square. The master square tests for gaps along an element; however, it cannot verify an ALL element test. It cannot detect whether the entire surface is within the control tolerance.

A similar test for angularity is an accurate angle measuring tool called a vernier bevel protractor. Here the protractor is not used to *measure* the angle of the element, but rather in geometric inspection, it is used to set up a master template to test individual elements similar to the master square.

Another common test for orientation is called an Optical Comparitor. This device projects enlarged versions of an object's profile onto a screen. The screen is able to rotate to exact angles and has a grid to interpret sizes. It is common to enlarge the object by 10, 15, or 20 times its normal size. The resultant image is easier to measure or test. However, similar to the templating method just discussed, optical comparitors verify only the profile of an object but cannot test an entire surface for all element compliance. Comparitors are also very useful in verifying geometric profile controls.

Challenge Problem 7–1

Answers follow.

Refer to the appendix drawing—Stabilizer Bracket.

1. How many controls of orientation do you find on the drawing?
2. Note that there is a multiple control involving both flatness and parallelism for the same surface. The flatness is a tighter control than the parallelism. Could the numbers be reversed? That is, could the parallelism be a closer control than the flatness? Explain.
3. Note that the position of the .312-inch diameter hole at the top of the wing is given as a distance up from datum -A- (3.507) and is also located on a second line. What is the relationship of this second line to datum -A-?
4. The inner surface of the entire wing must be parallel to which datum?

Answers to Challenge Problem 7–1

1. Five Controls: three parallelism, one angularity, and one perpendicularity control.
2. No. Parallelism embodies a control of flatness. A surface can be flatter than the parallelism because it could reside inside the parallel control zone. However, a surface cannot be less flat than the parallelism control because it would violate the parallelism control zone.
3. This second coordinating line must be at an angle of 40 degrees to datum -A- and within an angular tolerance zone of .010 inch.
4. Parallel to datum -C-.

Orientation Review

Orientation fits very well into our system of controls of part geometry. We have learned that:

CHAPTER 7 Controls of Orientation □ 87

- All three controls define a theoretical tolerance zone that is a pair of parallel planes/lines whose distance across is the orientation tolerance. All elements upon the controlled surface must lie within this zone. (Special case: the zone is a cylinder when controlling centerlines of holes.)
- Can be either a SINGLE or ALL element control. When using orientation to control surfaces, the entire surface must fit into the tolerance zone. If this extensive control is not required, the designer may consider a single element control. This is signified with a note. (ANSI Y14.5-M 6.6.1.1).
- All orientation controls refer to one or more datums.
- Each orientation control has a basic angle assigned relative to the control datum.

  90 = Perpendicular   0/180 = Parallel   Assigned = Angularity

- The basic angle has no tolerance. The tolerance zone is then built around this theoretical, basic angular model.
- Inspection of orientation is carried out using a standard layout table and indicators. Coordinate measuring machines (CMM) can also inspect orientation. Other tools and methods can verify orientation—for example, master squares or comparitors.
- Orientation is a control that can use a cylindrical tolerance zone when controlling the centerline of holes.

CHAPTER 8

Material Modifiers

We have studied all but four control characteristics thus far. It is now time to pause and learn a set of general concepts that apply to these controls—the concepts of material state modification. These are "Maximum Material Condition" (MMC), "Least Material Condition" (LMC), and a cancellation of material condition called "Regardless of Feature Size" (RFS).

The symbols are:

$$\text{MMC} = \text{\textcircled{M}} \quad \text{LMC} = \text{\textcircled{L}} \quad \text{RFS} = \text{\textcircled{S}}$$

One or more of these material condition modifiers may appear on the drawing inside a feature control frame. They affect only the individual control where they appear. If either MMC or LMC are shown, the tolerance stated in the control frame may be increased in actual practice. This increase, called "bonus tolerance," depends upon the machined size of the part.

The MMC or LMC symbol tells the shop that the tolerance stated on the drawing is a worst case, tightest fit tolerance. Almost always, the real parts as produced will have more tolerance than that which is shown in the frame. In fact, the designer never meant for that tolerance to apply except at the tightest extreme of fit or function. MMC means the most material permissible under the tolerance range, while LMC means just the opposite—the least material within the size tolerance.

Under the geometric system, the bonus tolerance is related to the feature size. As the size deviates from either MMC or LMC, extra tolerance is earned and added directly to the stated tolerance.

RFS Means No Bonus Tolerance

If the RFS symbol is shown, there is no bonus tolerance. The RFS symbol denotes that, according to the part function, there is no relationship between size and control tolerance. That which is stated on the drawing is all there is to use. This is no different than nongeometric tolerance; for example, the size and location of a hole. No matter what the hole size is, the position tolerance is a constant. We need little discussion about the RFS symbol; it simply means "what you see is what you get."

Understanding the MMC and LMC symbols and the concepts behind them is essential. A large portion of the calculation/manipulation skills you have yet to learn is based upon Material Condition.

Based upon Functional Fit

The geometric reason that we can have bonus tolerance is based upon the way parts assemble. We find in real functional assembly and fit that straightness, flatness, perpendicularity, parallelism, angularity, and position tolerances can be linked to the measured size of the feature to which they apply. If the actual size is not at MMC/LMC then the geometric tolerance can gain a bonus amount in direct proportion.

For example, in Figure 8-1, a hole in a part must be closely dimensioned and toleranced for location, so that it will assemble over a second part with a pin protruding. The engineer must take into account the fit at the tightest possible state of the hole. The position tolerance of .005 inch was designed to fit correctly at MMC. We know that because of the MMC symbol in the frame. What is the MMC size of the hole? (Hint—the tightest fit case would be the smallest hole.)

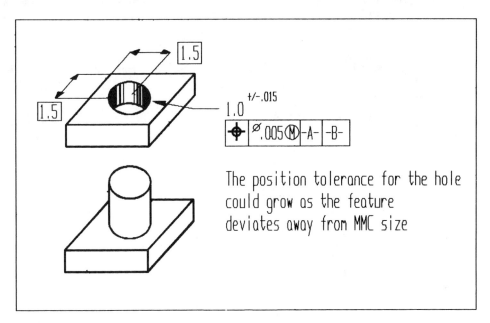

Figure 8-1 Bonus tolerance is based on functional assembly.

> **Student Note:** We have not studied geometric position yet, but position is the best example to illustrate MMC bonus. Geometric position for a hole locates a perfect position then tells the user the diameter of the target zone around this perfect position. The centerpoint of the hole must fall into this zone. A more thorough discussion is found in Chapter 10.

In Figure 8-1, what then is the MMC size of the hole? What is the size that is the tightest fit over the pin?

$$1.00 - .015 = .985$$

The MMC size is .985 inch because that is the tightest fit permissible when the basic size and the extreme of tolerance are considered. Therefore, the position tolerance of .005 inch applies at the MMC size. You know this because of the MMC symbol in the frame.

Assume that the hole is made .010 inch bigger than MMC. Could the hole then be less accurately located and the parts still assemble? Yes. As the hole gets larger it needs less accurate location to assemble with the pin. How much location tolerance did we gain—exactly .010 inch. With the MMC symbol present to give us permission, we gained a bonus of .010 inch bonus location tolerance. A hole size of .995 has an actual working tolerance of:

$$.005 \text{ (stated tolerance)} + .010 \text{ (bonus)} = .015 \text{ inch}$$

In this text, this full working tolerance is called the "Earned Tolerance." The earned tolerance is always related to a part already machined—it depends on an actual feature size.

In this example, the position tolerance is geometrically linked to the size of the hole. As the hole size changes away from MMC size, the location can grow in direct proportion.

Please be aware, we will study more about calculating bonus tolerance and position in Chapter 11. This chapter is needed to show you how MMC and LMC apply to the control characteristics we have thus far learned. They are the design flags that tell the shop when and when not to compute bonus tolerances.

Six Geometric Tolerances Can Be Linked to Sizes

In the geometric system we find that there can be a connection between the geometric size of a feature and its control tolerance. Six different geometric tolerances can grow as the part deviates from a given material condition—MMC or LMC, whichever applies.

Relationships exist between size and:

— Straightness
— Flatness
— Perpendicularity
— Parallelism
— Angularity
— Position

Applies To Centerlines Only

In every case, the control relates to the *Centerline* of the feature of size. Based upon the part function and with the symbolic permission on the drawing, as the features actual measured size deviates from either MMC or LMC, one of the above six characteristics can grow. This concept applies to flat features that have centerplanes as well as round features. Examples of features with centerplanes are tabs, bosses, and slots. These all have width and are centered around a central line. Any of the above six controls that relate to the centerline/plane of a feature can be modified (that is, they can gain a bonus tolerance) if the function allows.

For example, the straightness of the centerline of the pin in Figure 8-2 could grow (be less straight) if the pin was smaller than MMC size. This makes functional sense. Suppose the pin was to fit into a hole. The straightness tolerance of .002 inch was calculated at MMC—we know this from the MMC symbol in the frame. As the actual pin diameter then deviates from MMC—as it becomes smaller in diameter—it could be a little more curved and still fit.

In Figure 8-2, the centerline of the pin is to be straight within .002 inch. You notice that the MMC symbol is present. That means the .002-inch tolerance is the

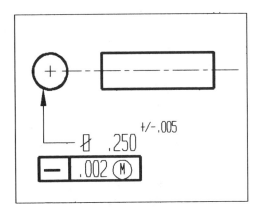

Figure 8-2 A bonus straightness tolerance may be applied to the centerline of this pin.

worst news possible—that the .002-inch straightness tolerance for the centerline was calculated when the pin was its maximum material condition size, .25 + .005 = .255 diameter. If the machined size is something less than .255", then the tolerance for centerline straightness can grow in direct proportion.

For example, you measure the pin at .249 inch. It has deviated from MMC margin size by .006 inch. This .006 can be added directly to the stated tolerance of .002".

For a .249" pin, the straightness tolerance is .002 + .006 = .008 inch.
Other machined pin sizes will have different total tolerances.

— We call the tolerance shown in the frame the *Stated Tolerance*
— The name of the .006 is called the *Bonus Tolerance*
— The resulting sum of .008 inch is called *Earned Tolerance*

Remember this earned tolerance can only occur when we are dealing with real machined parts. The bonus is gained by measuring a resultant size and then comparing this size to the MMC/LMC size. The difference in actual and MMC/LMC size (the bonus) is then added to the stated tolerance.

Absolute Value of the Difference in Two Numbers

When we subtract one quantity from another, we say this is the difference in the two numbers. We will be discussing the difference in numbers throughout this book. The order of the two numbers will change depending upon whether the feature is an outside feature, such as a tab, or an inside feature, such as a slot, and also upon whether the control is MMC or LMC. The resultant difference will always be a real number (the bonus tolerance) needed to apply to a problem—thus it will be a positive value.

It does not matter in what order you do the subtraction; the numbers without regard to sign value (+ or −) will be the same—for example, 13 − 5 = 8 and 5 − 13 = −8.

When we are interested only in the positive number, it is called the absolute value. This is symbolized by the placement of vertical lines around the problem—both, $|13 - 5| = 8$ and $|5 - 13| = -8$

The second difference statement is spoken as "Absolute value of five minus thirteen equals minus eight."

The Material Conditions

Since the entire process starts with the two material conditions, MMC and LMC, let's define them further. Learn them well. Although they seem simple, predictably many of your early learning errors will be due to miscalculating the MMC or LMC size.

Maximum Material Condition—Concept #17 Ⓜ

MMC occurs when a certain feature has the "Most Material" permitted by the print size and its associated tolerance. For the pin in Figure 8-2, the design size is .250 inch and the most metal permitted is another .005 inch. Therefore, MMC is .255 inch—*the most metal.*

Avoid saying "the biggest size" because some features are at MMC when they are their smallest size. For example, a part with a slot cut in it actually has more metal when the slot is the smallest possible. In this case the MMC size for the feature was the smallest version but had the most metal. Always think "most metal" before making the actual calculation.

MMC is used as a basis for assembly. When designing assembly parts, such as the hole fitting over a pin example above, the MMC or LMC size is a prime factor in determining the position tolerance. This is the worst case—this is where the location tolerance is born. Other factors come into play as well, such as straightness and perpendicularity of the hole. When all of these variables are taken into account for the hole, this is called the "virtual condition" of the hole. This is the bottom line for fit—the absolute worst case. Any size deviation away from virtual condition then makes assembly more possible.

Stop and think before you determine the material condition size. It will always be at one extreme or the other of the possible size range for the feature; in other words, all the tolerance either added or subtracted from the design size. Think in terms of material on the part!

Least Material Condition—Concept #18 Ⓛ

LMC is just the opposite of MMC. It is the state where the feature has the least material permitted by size and tolerance. LMC is used primarily to modify feature positions to maintain minimum thicknesses of walls, such as the thickness of a hydraulic tube wall or a bearing cap. In Figure 8-3, LMC could be used to control the thickness of the walls on this mounting block.

When calculating the bonus tolerance for a control LMC, other than the material condition, nothing changes the process. Compute the LMC size, then measure the actual size and find the deviation which is bonus tolerance. The bonus will be the difference between the LMC and Actual sizes. Add this bonus to the stated tolerance.

Challenge Problem 8-1

Solve for the MMC or LMC size.

94 ☐ Unit II The Working Geometric Concepts

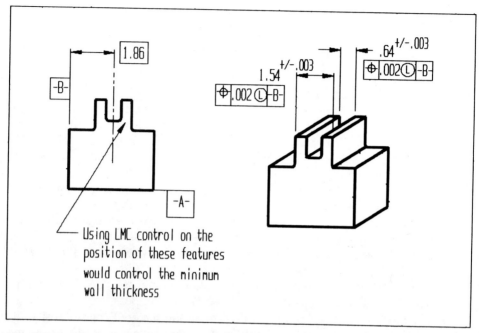

Figure 8–3 As the slot or boss deviate away from LMC size, their position tolerance will grow in direct proportion.

Answers follow.

1. A block has a design height of 3.000 inches and a size tolerance of +/− .002 inches.
   A) What is the MMC size for the block?
   B) What is the LMC size?
2. Return to Figure 7–3. For the .50 hole
   A) What is the MMC size?
   B) What is the LMC size?
3. A machined bushing has a dimensioned inside diameter of 1.375 inches. It has a tolerance on the inside diameter of +.010 and −.002 inches. What is the LMC size for the hole?
4. What is the LMC size for Figure 8–2?
5. The tolerance is +/− .003" for both dimensions in Figure 8–3.
   A) What is MMC for the .64 slot?
   B) What is LMC for the 1.54" boss?

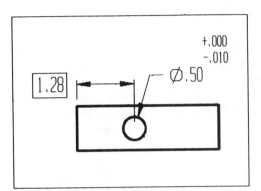

Figure 8-4 What are the LMC and MMC sizes of this feature?

6. For Figure 8-4
   A) What is the LMC size of the hole?
   B) What is the MMC size?

Answers to Challenge Problem 8-1

Answers are all in inches.

1. A) MMC = 3.002"
   B) LMC = 2.998"
2. A) MMC = .497"
   MMC is the smallest hole.
   The part has the MOST metal when the hole is at its smallest.
   B) LMC = .500 + .003 = .5003 inches
3. Remember, LMC means the least material. This is a hole, the least material on this part considering the hole is the biggest hole possible!
   1.375 + .010 = 1.385-inch LMC diameter.
4. LMC is .245 inch.
5. A) MMC = .640 −.003 = .637 inch
   B) LMC = 1.540 −.003 = 1.537 inch
6. A) LMC = .500
   Tolerance is unilateral—design size is LMC size.
   B) MMC = .490

The Earned Tolerance Calculation Procedure—Concept #19

In practice, the process for calculating the earned tolerance is not complex. It will always follow the same procedure. We will make an MMC computation for Figure 8-5.

Note, earned bonus problems must be based upon real parts; therefore, for all exercises in this text involving MMC/LMC earned bonus, *the proposed actual measured size of any print requirement will appear in brackets next to the given dimension.* In this case, the actual part produced measures .621 inch. Using brackets to tell you the measured size will be standard throughout the book. Base your bonus calculations on this measured size.

Problem Example Figure 8-5.

What is the earned perpendicularity tolerance of a part produced with a .621-inch diameter hole? Because of the MMC symbol, we know that the given .003-inch perpendicularity tolerance only applies to the hole centerline when it is at its MMC condition.

1. From the print find the MMC size.
   Determine the material condition limit size of the hole.

   MMC Size = .625−.005 = .620

2. Measure the part to determine actual size deviation.
   We determine the part is (.621) actual size.

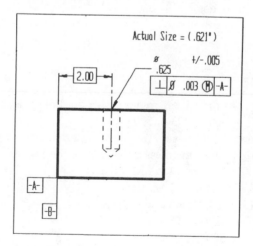

Figure 8-5 Compute the MMC size.

3. Calculate bonus tolerance.
   This is the absolute difference in actual and MMC sizes.

   $|.621-.620| = .001$-inch bonus

4. Compute earned tolerance.
   Add bonus tolerance to given print tolerance.
   (The perpendicularity tolerance is .003 on the print.)

   $.003 + .001 = .004$

For this example the earned tolerance is .004 inch. This is for a part whose hole diameter is .621 inch. Other hole diameters would have other earned tolerances.

The four-step procedure to calculate earned tolerance is:
1. Find MMC (or LMC) size.
2. Measure actual size.
3. Find the absolute difference—subtract to get bonus.
4. Add the bonus to stated tolerance on the print.

Challenge Problem 8-2

Instructions:
   The following problems involve calculating the earned tolerance. They will apply only to those geometric characteristics that can have bonus tolerance that we have thus far learned: Straightness, Flatness, Perpendicularity, Angularity, and Parallelism. Remember, the actual feature size for the question will be found in brackets on the drawing.
   Please label each step in your computations—this will train you for industrial work where more steps will be involved. The answers follow the exercises.
   1. Recompute the earned tolerance for Figure 8-5 for the following actual sizes.
      A) Actual size = (.624)
      B) Actual size = (.627)
      C) Actual size = (.630)
   2. Find the earned parallel tolerance for Figure 8-6.
   3. For Figure 8-7, find the earned centerline perpendicularity tolerance for the slot width shown?
   4. Find the earned angularity tolerance for Figure 8-8 if the hole measures .8115 inch.
   5. Determine the earned straightness tolerance for Figure 8-9.

98 □ Unit II The Working Geometric Concepts

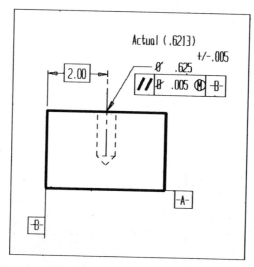

Figure 8-6   Compute the tolerance.

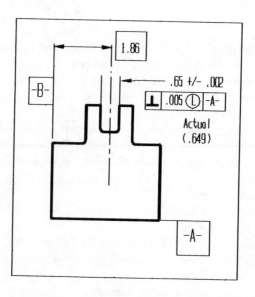

Figure 8-7   Find the perpendicularity tolerance for this slot as machined.

CHAPTER 8 Material Modifiers □ 99

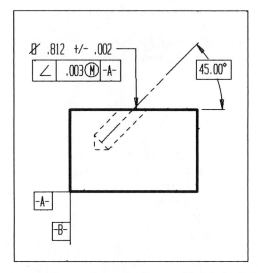

Figure 8–8 Find the angularity tolerance after machining the hole to .8115 inch.

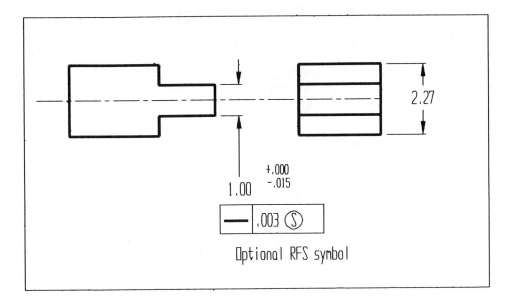

Figure 8–9 The RFS symbol indicates *no bonus tolerance*. The control remains RFS if no symbol is present.

Answers to Challenge Problem 8–2

1. The recomputed earned tolerances for Figure 8–5 are:
	Actual Size	Earned Tolerance
A)	.624	= .007
B)	.627	= .010
C)	.630	= .013

2. Earned parallel tolerance = .0063 inch
3. Earned perpendicularity tolerance = .008 inch
   Remember, this is an LMC problem!
4. The earned angularity tolerance = .0045
   MMC = .810"   Actual = .8115   Bonus = .0015
   Earned angularity equals bonus plus stated tolerance
   .0015 + .003 = .0045
5. .003 inch—the RFS symbol is present! NO BONUS! Note: the MMC bonus could apply to this detail drawing but the function of the part doesn't allow the extra out of straightness. The designer has placed the RFS symbol in the control frame—there is no bonus straightness tolerance for any feature size. See the next section for more information.

Regardless of Feature Size —Concept #20 Ⓢ

No Bonus Tolerance

The Regardless of Feature Size (RFS) symbol in the frame tells the user that there is NO BONUS tolerance. The symbol for this is a circle with an "S" inside. When this symbol is present, there is no bonus tolerance. The stated tolerance is the limit.

The best way to understand RFS is that you are simply working with a size tolerance and a control tolerance and they just do not change. This is what you have always done on nongeometric parts. There are two general rules concerning material condition symbols.

Geometric Rule #3—RFS is Automatic for ALL Controls other than Position

For all the controls we have thus far studied—in fact, all the controls other than position—if there is no material condition modifier, then RFS applies.

So, referring to Figure 8-8, if there was no MMC symbol as the hole deviated from MMC, the angularity tolerance would be the same as that shown in the original frame. The designer may choose to put the RFS symbol in the frame but without it, as in Figure 8-10, RFS still applies. Twelve of the thirteen controls are automatically RFS.

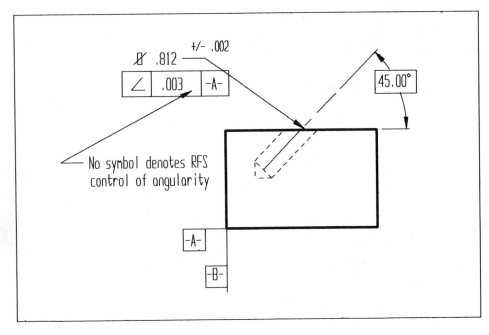

Figure 8–10 This centerline angularity control has no bonus tolerance.

Rule #3 after 1982 Standards

Per the ANSI standard, *for controls of position the designer must specify either MMC, LMC, or RFS*. If there is no material condition symbol and the drawing is dated after 1982, the drawing is incorrect, but see the next paragraph. Occasionally we encounter drawings of this nature.

Rule #3 before 1982

Previous to 1982, MMC was considered automatic with position only. Under the pre-1982 standard, the designer had a second option of placing the MMC symbol in the control frame. The 1982 modification to the ANSI Y14.5 became Y14.5-M and cleared up this ambiguity.

Before 1982, if the designer wished RFS to apply, he/she had to specify it; however, with no symbol shown, MMC could be assumed. There was no LMC symbol previous to 1982.

This understanding of history is important for manufacturing people because often we are making parts for replacement or that have not been redrawn since the 1982 standards which means we are working to older design standards under the Y14.5 rules. You might have bonus tolerance on position controls even though the drawing has no symbol. It was common practice to assume automatic bonus tolerance pre-1982.

Review of Material Condition Modifiers

- The symbols are: MMC= Ⓜ LMC= Ⓛ RFS= Ⓢ.
- These symbols appear in feature control frames.
- They affect only the individual control where they appear.
- If either MMC or LMC is shown, the tolerance in the frame may be increased in actual practice.
- This increase called "bonus tolerance" depends upon the machined size of the part.
- If the RFS symbol is shown, there is no bonus tolerance.
- Material condition bonuses are based upon functional fit. That is, if a part feature size changes, the control tolerance can grow in direct proportion to the size change away from either MMC or LMC.
- Based upon function, bonus tolerance relationships can exist between size and Straightness, Flatness, Perpendicularity, Parallelism, Angularity, and Position.
- Applies to feature Centerlines or Centerplanes only.
- MMC occurs when a feature has the "MOST MATERIAL" permitted by the print size and its associated tolerance.
- LMC is the state where the feature has the least material permitted by size and tolerance.
- LMC is used primarily to control minimum thicknesses of walls.
- The four-step procedure to calculate earned tolerance:
    1. Find MMC/LMC size
    2. Measure actual size
    3. Find the absolute difference—subtract to get bonus
    4. Add the bonus to stated tolerance
- The RFS symbol in the frame tells the user there is No Bonus.
- RFS is automatic for ALL controls other than position.
- For controls of position the designer must specify either MMC, LMC, or RFS (post-1982).

CHAPTER 9

Characteristic Controls of Runout

Runout controls the surface of a rotating part as compared to a central datum. Runout deals with objects that rotate and must have a specified minimum surface wobble. Runout controls the distance of a surface element from a central axis therefore runout is always with reference to a datum. The rotation axis is established by one or more datum features of the part.

In simpler terms, we spin the part 360 degrees about an axis and test it with an indicator for surface wobble. The full indicator movement (FIM) must be less than or equal to the control tolerance.

There are two versions of runout:

1. Circular runout—A SINGLE element, two-dimensional control.
2. Total runout—An ALL element, three-dimensional control of a surface.

Circular Runout—Concept #21

A Single Element Control

The single arrow symbol tells the user that this is to be an individual element control. The engineering function is not to control surface relationships but rather that each element does not wobble any more than a given tolerance.

On the part in Figure 9-1, circular runout may be applied to surfaces A, B, C, and D—a straight diameter, an angular surface, a contoured surface, and a 90-degree face. In each case, when checking the runout as in Figure 9-2, the test is at 90 degrees to the surface at the point of contact. In each case, the inspector would touch the indicator to several elements and if each had no more FIM than the control tolerance, the part is acceptable. Notice in Figure 9-1, surface A, that no two elements would be the same diameter yet each must have no more runout than .003 inch FIM.

On Figure 9-1, all surfaces (except surface D), runout also controls roundness. While the indicator test does not determine what the actual roundness is, it

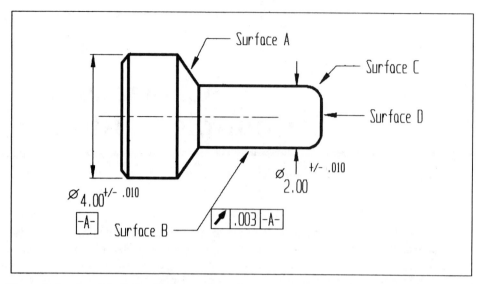

Figure 9–1 Control of circular runout.

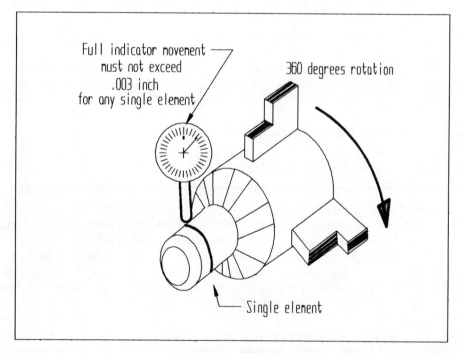

Figure 9–2 Circular runout is tested by rotating the controlled feature about a datum axis.

does tell the inspector that the roundness of the element is equal to or better than the FIM of the individual test element.

Recall in Chapter 6 that indicating the part while spinning determines runout but not actual roundness. However, if the FIM is within the roundness control limit, then one may correctly conclude that the roundness is no "worse than" the tested FIM.

The element tested might be perfectly round yet be off center to the reference datum. However, the FIM will be the upper limit of the actual out-of-roundness.

On surface D in Figure 9-1, there is embodied a control of single element flatness, but only for an individual element. All other elements need not compare to this single element. Each element test must only be within the control FIM.

Inspecting Runout

This is a straightforward process. While in contact with an indicator, the part is rotated 360 degrees around a central datum axis. The full indicator movement must be less than or equal to the control tolerance. The number of tests taken is arbitrary. The more elements that you test, the more accurate the result. If any single element fails (exceeds the tolerance FIM), then the part is out of geometric runout tolerance. Any feature relating to a central axis of rotation may be controlled with circular runout.

Corner radii, grooves, and threads can be controlled. When dealing with threads or gears, the pitch diameter is the test point. The control applies to the pitch diameter (see ANSI Y14.5-M 6.7). Each element is an individual test—circular runout does not control an entire surface.

Total Circular Runout—Concept #22

In Figure 9-3, the double arrow symbol, total runout, is an ALL element control. The three-dimensional tolerance zone is a perfect model of the surface with respect to a datum axis. When rotated about the datum axis, the FIM for the entire surface must not exceed the control tolerance. The part is repeatedly rotated through 360 degrees and the composite of all elements is compared. The difference in the lowest reading on the surface and the highest reading must not exceed the total runout FIM control.

Pre-1982 Symbol

The double arrow symbol was instituted at the 1982 update to ANSI Y-15.5-M. This further symbolized the system and eliminated ambiguities. Previously, the symbol was a single arrow, the same as circular runout, except the word "TOTAL" was printed just below the symbol.

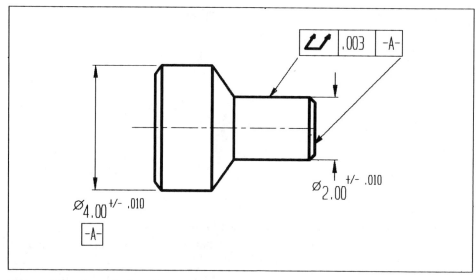

Figure 9-3 Control of total runout.

Inspecting Total Circular Runout

This control is relatively easy to inspect. The part is rotated and an indicator is swept across ALL elements—the entire controlled surface. The lowest point is compared to the highest point, the difference being the total FIM. This must be less than or equal to the control tolerance. This would be .003 inch in Figure 9-4.

Types of Surfaces Controlled by Runout

On the part in Figure 9-1, only surfaces B and D are defined in ANSI as available for total runout control. This is due to a technical complication in inspecting the angular and contoured features. Previously, when inspecting these features, the difficulty would be that the *tip* of the indicator must be moved along the perfect definition of the surface being controlled.

Although undefined by ANSI, controlling total runout on contoured surfaces is possible with modern computer equipment. The model, perfect surface would be the template for the test indicator movement. Inspection of contoured surfaces for runout would involve moving an indicator probe tip along the model surface contour while keeping the indicator normal to the surface.

Runout Always Applies RFS

Since runout is a control of surface elements to minimize wobble of spinning

CHAPTER 9 Characteristic Controls of Runout □ 107

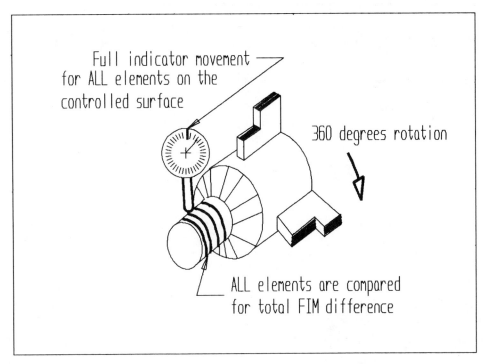

Figure 9-4 Inspecting total runout.

objects, it always applies regardless of feature size. It matters not whether the spinning is at MMC or LMC or between, the wobble must be controlled. Both circular and total runout apply RFS.

Review of Runout

- Runout is a control of surface wobble on rotating parts.
- Runout always refers to a datum axis for rotation.
- Runout may be an individual or ALL element control depending upon the functional requirements. Total runout can be fractionally more expensive to achieve.
- Surface elements are tested at 90 degrees to the surface.
- Runout is evaluated by rotating the part around the datum axis. The FIM is noted either for single elements, as with circular runout, or the entire FIM composite is compared for total runout.

Special Note. Embodied within total runout are implied controls of straightness, roundness, and coaxiality. In effect, runout also can control form (ANSI Y14.5-M 6.7.2.2).

In Chapter 10, we will explore a similar control. We will examine concentricity and then compare concentricity to runout. This will help your understanding of both controls. Both controls deal with co-axiality—the comparison of features to an axis.

CHAPTER 10

Characteristics of Location

Our final grouping, location, includes two controls—Concentricity and Position. In this chapter we will learn a great deal about the fun part of geometrics, the manipulative skills. Position forms such a strong central theme in geometrics that in years past, the entire subject was referred to as "true positioning." Today we call it geometric dimensioning and tolerancing but, as you will soon see, many of the deep skills within this subject are based upon position and material conditions.

Both concentricity and position control the centerline of a feature within a tolerance zone built around the perfect location. The perfect location is designated from one or more datums. As with all the other geometric controls, location defines a perfect model—in this case the perfect target point, centerline, or target center plane. The location control is built around this perfect target.

Actual drawing location dimensions for the perfect target are given as basic—without tolerance—perfect models. There is no tolerance for the perfect location—the tolerance is built around this model.

Concentricity—Concept #23

Verifies Coaxiality

The geometric control of concentricity verifies coaxiality of part features. That is to say the centerline of one feature is controlled to be within a cylindrical tolerance zone built around the perfect center of another reference feature. The reference feature must then establish a datum.

In Figure 10-1, datum -A- is established by the 2.0 diameter. The centerline of the 1.0 diameter is to be within a cylindrical tolerance zone, .003 inch in diameter, built around datum -A-.

The word "concentric" means "having the same centers." The definition of

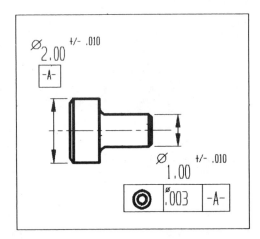

Figure 10-1 Control of concentricity.

coaxiality is similar—"sharing the same centers." Concentricity verifies and controls coaxiality. Concentricity is not the only control that can verify coaxiality.

The term coaxiality is the condition whereby two centerline axes are coincidental (parallel and occupy the same space). Coaxiality may also be controlled with runout and true position. We will make a comparison shortly.

Inspection / Engineering Note. Concentricity is difficult to inspect and is specialized. *Either runout or position should be used to control coaxiality, whenever possible.*

The Function Of Concentricity Is To Control Vibration On Spinning Features

To further understand concentricity, consider the major function. The major engineering purpose of concentricity is to control vibration of an object that rotates rapidly. Concentricity controls the center of mass of a dynamically rotating feature with respect to a central axis.

The difference between concentricity and runout can be demonstrated (see Figure 10-2). The surface of the head of this bolt has a great deal of runout. While rotating, an indicator on the bolt head surface would show a large FIM. However, the bolt head is concentric to the bolt, is it not? It is possible to spin this bolt on its long axis and not have any vibration.

Concentricity is a specialized control and has two limitations that can become costly: it is difficult to verify and concentricity always applies RFS, no

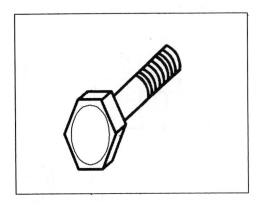

Figure 10-2 This bolt head is concentric to the bolt.

bonus tolerance. Concentricity is often misused for controlling round objects that do not spin nor do not need dynamic balance.

An example would be locating a counterbore for a bolt head around a bolt hole—concentricity is the wrong control for this function. This in turn drives up costs and limits bonus tolerancing. Only when the object revolves and vibration is a concern should concentricity be used.

First, concentricity is *always* RFS because it deals with centers of mass. There can be no bonus tolerance using concentricity. While an object might vibrate less as it deviates from MMC, not much change would occur. Functionally, concentricity is an RFS control.

When controlling coaxiality, runout and position have advantages over concentricity. Runout is much easier to inspect than concentricity and position can be applied MMC or LMC thus a bonus tolerance is possible.

Consider Figure 10-3. This counterbored plate is designed to nest the bolt of Figure 10-2. Functionally, as the counterbore becomes larger (deviates away from MMC size), then it could be less accurately located with respect to the bolt hole. This bonus tolerance is not possible with concentricity.

If the large counterbore is at MMC size, the bolt will fit in. As the counterbore deviates from MMC, should not the coaxiality be able to increase? Yes, it should. But by using concentricity to control this feature, the designer has prevented this bonus. The location of the counterbore with respect to datum -D- (the hole) should be controlled using Position MMC.

Inspecting Concentricity

The second drawback is in inspecting concentricity. Concentricity is used to control the center of mass or center of revolution of a part feature. Therefore, when

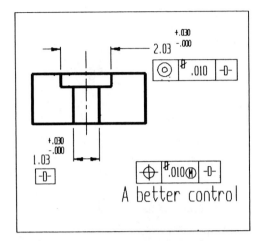

Figure 10-3 Changing the concentricity to position will allow bonus tolerance and better serve the design function.

inspecting concentricity, the surface of the controlled feature must be dealt with as with establishing a datum. Only after the surface is contained may we extrapolate the true center of the controlled object. This centerline establishment process can become difficult.

Example A.

Inspect the concentricity in Figure 10-1. Suppose we hold the part on the 2.0" datum -A- in a precision collet. This then establishes the controlling datum -A-. Now we rotate the part and place an indicator against the surface of the 1.00 diameter. This would produce an indicator FIM. Let's say you read .010. Does this mean that the center of the 1.0 diameter is within a .010 geometric concentricity?

It could if the part was round. The problem with this test is that we are really trying to verify where the centerline of the controlled surface lies by indicating the surface. We are testing surface elements with probable irregularities. This test is valid but, keep in mind, it shows the part may be within tolerance but does not show the actual concentricity.

In order to be sure, the 1.00 diameter must be treated as a datum. A possible method would be a ring gage over the tested diameter. This perfect cylinder slipped over the test surface would average out the irregularities and the center of the gage would then be compared to datum -A-. This would require many different ring gages in stock.

CMM and Concentricity

A computer inspection machine can verify concentricity by testing several circular elements. Averaging out each to the best fit circle, then connecting the center points of each determines the true centerline of the controlled feature. CMM's eliminate much of the difficulty in inspecting concentricity.

Example B.

Consider Figure 10-2. Suppose we wish to control the concentricity of the bolt head. Runout would be easy to inspect: we would simply hold the bolt in a collet, place an indicator on the bolt head, then rotate the part. As each high point passed by the indicator probe, it would be recorded. Each high point would be compared for FIM. We were testing surface elements. This could also be done to the low points as well.

Inspecting the concentricity of the bolt head becomes a real problem. We must deal with the entire shape of the hex head—all the surface shape and irregularities (treat the bolt head as a datum)—because the center of the entire mass is in question. Somehow, we must then establish where true center of the hex head is.

Could it be accomplished with a ring gage that just slips over outside diameter of the hex head? What about the irregularities of the flats? This circular contact would not take into account the flats and their irregularities. Would they cause dynamic balance problems? The entire shape must be considered.

The *theoretical* solution would be a perfect hexagonal ring gage that can be shrunk down to fit the hex bolt (Figure 10-4). It would average out all the surface irregularities of the hex head and establish the center of the hex. The center of this imaginary hex ring gage could then be tested against the bolt's centerline and we would have verified the concentricity. In practice, this is impossible.

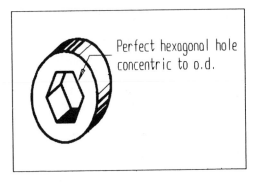

Figure 10-4 An imaginary hex ring gage to solve the concentricity inspection.

How would you inspect the hex head? There are two practical solutions, both involving modern technology. One is a computer profile testing machine. A single element could be tested as the part is spun. The computer could then analyze the hexagonal element and determine the true center of the cross section. Testing several elements would then establish the true center of the hex.

A second and less accurate solution would be an optical comparator where templates could be drawn and placed over the projected hex image on the screen. The part could be rotated and each flat and high point of the hex compared to the template. But this test has the limitation of being able to "see" the entire surface of the hex head. Optical comparators project profiles onto a screen. Thus they are limited to single element tests similar to you looking at the shadow of the hex head.

From these two examples, you can see that either position or runout would be acceptable and far easier to inspect for control of the coaxiality of the hex head. So when we do have concentricity, how do we inspect it?

First, keep in mind that, assuming uniform material density throughout the feature in question, your objective is to find the center of mass of the controlled feature with respect to a central datum axis of rotation. This can be done by testing the physical surface and deriving the true center.

The standard indicator test is most common and can be valid. It can detect compliance to the tolerance, but the actual value of the eccentricity can be difficult to derive. Eccentricity is the amount of deviation from perfect concentricity—the opposite of concentric.

Let's look at some examples and answer some questions. Visualize an object with two different diameters on a single shaft. The first diameter is the control datum feature -A- while the second diameter must fall within a concentricity tolerance of .010 with respect to the first features axis.

Test 1.

Holding the part in an inspection quality collet or other device that establishes the center of datum feature -A-, rotate the part while testing several elements along the controlled surface. Suppose you get .009 inch FIM for several elements, and the highest readings are found on the same general side of the part for each element. What does this tell you? Mainly, this result tells you that the feature is within the concentricity tolerance. The value of concentricity is near to .009 inch but could actually be less than the FIM shown. Why? Some of this may be out of roundness. You actually have not averaged out the center of each element—you have tested runout.

Test 2.

Holding the datum feature in an inspection quality collet, test several elements for FIM. The FIM is .012 inch, with two high points and two low points per element test. The high points appear to be opposite each other and the low points as well. All points appear to be in the same general position on the tested feature. Is this part out of tolerance? Probably not. What is the problem?

It appears to be an oval and might actually be exactly on center. It could be spun and have absolutely no vibration occur. Recall the bolt head test. There would have been lots of runout yet the concentricity was OK.

The problem is that simply looking at the indicator FIM shows you only the surface runout. The true center of the controlled surface is a derived quantity.

Since this surface would seem to be on center but oval in shape, more sophisticated testing is required for this part; it might actually be perfectly concentric, but not round. Can you see why, in this case, the concentricity is dependant upon the outside form—the roundness of the controlled feature?

Test your understanding.

Case A.

Could this part just discussed be designed with a roundness tolerance of .010 inch on the controlled feature and a concentricity tolerance of .005 inch on the same feature?

Case B.

Could it have a roundness of .005 inch and a concentricity of .010 inch?

Answers

Case A.

Yes, it could.

Although, as in the case of the perfectly centered oval, it is possible for the un-round feature to be concentric, it is highly unlikely. Additionally, the part would be nearly impossible to inspect. Finding the true center would be a mathematical nightmare.

Case B.

Yes, it could.

Truly, only a computer can test for the absolute value of eccentricity. However, if the above indicator–FIM test does not exceed the control tolerance, you may say for certain that the part feature is within concentricity tolerance—but that is all! The actual value is a computed number dependant upon establishing the average centerline of the controlled feature then comparing it to the datum axis.

Review of Concentricity

- Concentricity is a control of location.
- A perfect centerline model is defined by a control datum feature.
- The tolerance is built around the perfect model centerline.
- The tolerance is the distance across the cylindrical zone.
- The geometric control of concentricity verifies coaxiality.
- Concentric means having the same centers. Coaxiality is similar—"sharing the same centers."
- Concentricity controls vibration on spinning features.
- Concentricity is a specialized control having two limitations.
  1) RFS always.
  2) Inspection can be difficult.
- If the tested element is round, testing for concentricity also tests roundness as well. The roundness of any element will not exceed the FIM and directly affects the result of the concentricity test.

Control of Position—Concept #24

Position is perhaps the most interesting control in the system. In this part of Chapter 10, after you learn the basics of geometric position, the real fun part begins—the calculating skills. Then, in Unit III, we will explore in-depth manipulative possibilities that occur with geometric position.

Position defines the perfect location for the centerline or centerplane of a feature. The centerline of the controlled feature must then fall within a tolerance zone built around this perfect location.

In the case of a round feature such as a hole, a pin, or a counterbore, the perfect model location will be a centerline defined by two basic distances and/or orientations from datums. This centerline defines the true axis of the hole (Figure 10-5).

The most common shape for the tolerance zone will be a cylinder built around this perfect location line. The real centerline of the controlled feature must then fall within this tolerance zone. The diameter of the control zone cylinder is the geometric tolerance. Soon, we will see that the MMC and LMC bonus often apply to the size of this zone.

The perfect location model for a feature such as a tab or slot would be a theoretical flat centerplane surface. This perfect location for the centerplane of the feature is defined by a single basic distance and orientations from datums. The tolerance zone would then be a pair of theoretical parallel surfaces built around the perfect position. The centerline of the feature must then fall within this tolerance zone.

CHAPTER 10 Characteristics of Location □ 117

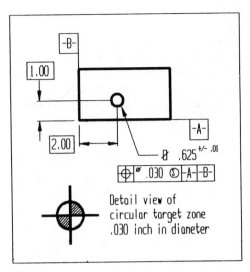

Figure 10-5 A geometric tolerance zone for the location of a hole centerline.

Centerlines or Centerplanes Only

Position controls only features with centerlines or centerplanes. If the feature has no centerline then position may not be used. For example, geometric positioning can not be used to locate a surface relative to another surface.

Always Refers to a Datum

Since a position is relative to some reference, it must always refer to a datum or datums. Position models will be located using basic distances from datums and possibly using controls of orientation as well. For example, the center axis of a hole must be 2.0 inches basic from a datum and parallel to another datum within .001 inch.

Material Condition Modifiers May Apply

Position must be either MMC, LMC, or RFS, but none are automatic. Per the 1982 standards, the designer is required to state either LMC, MMC, or RFS in the control frame when using position.

Pre-1982

Recall that previous to 1982, MMC was automatic for controls of position only. When there was no symbol, for position only, the user assumed that MMC was in effect. Also, there was no LMC symbol.

Two-dimensional Note

For the purpose of learning, in Unit II we will discuss only the two-dimensional aspect of position. This means that the tolerance zones for holes will then be circles instead of cylinders and for features that would have flat surfaces such as a tab or slot, the zone will be parallel lines rather than parallel surfaces. Three dimensionality becomes important during design and inspection, but for the needs of learning concepts, we will withhold three-dimensional aspects.

Basic Position of a Hole (Concept #25)

In Figure 10–5, the model position is defined with basic dimensions from datums -A- and -B-. This is symbolized by the placement of the box around the dimensions. (Pre 1982, we would see no box and the letters BSC or the word BASIC beside the dimension.) Either way, it means a perfect dimension around which the tolerance will be developed.

In Figure 10–5, the tolerance zone is a circle .030 inch in diameter. If the center of the produced hole falls within this control circle, the part is acceptable. For this example the control is RFS. We will explore material conditions as they apply to position later.

The round geometric control zone is an advantage over nongeometric tolerancing. Nongeometric tolerancing is also referred to as coordinate or rectangular dimensioning. We will compare coordinate dimensioning to geometric dimensioning.

Geometric Target Zone

In Figure 10–5, we see an exploded view of a geometric tolerance zone. Note that the center point of the controlled hole must fall anywhere within the .030 diameter circle. That means the center point of the hole may deviate .015 inch in *any* direction so long as it falls within the given .030 circle.

This circular target zone fits our concept of functional fit. Think about a pin in a hole with .015 inch clearance per side. The pin can move sideways inside the hole. Take, for example, a .970 pin in a 1.00 diameter hole. The pin can move .015 inch in any direction. Most importantly—*any direction*. If you plotted all the possible positions of the center of the pin, what would it describe? Answer—a .030 diameter circle centered around the perfect center point location. That is the functional fit of a round object and the natural positional tolerance for this part.

Coordinate Tolerance Zones Are Smaller

Now, let's see how that compares to a nongeometric coordinate tolerance. In Figure 10–6, a further expanded view of this .030 tolerance zone circle is shown. Here we see a corresponding nongeometric, rectangular tolerance zone compared

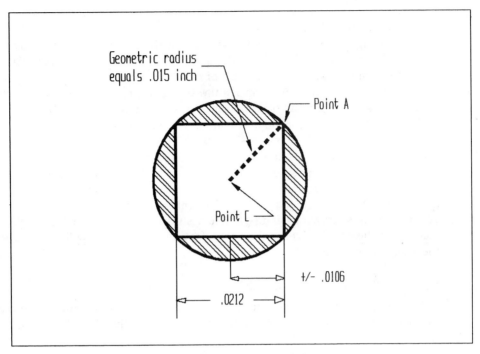

Figure 10-6 A .030-diameter geometric tolerance zone compared to a rectangular (coordinate) zone.

to the .030 geometric zone. The corresponding rectangular zone is much smaller. It will be given as a plus and minus tolerance for the locating dimensions.

Note that Point A is a valid location for the center of the controlled hole using either geometric or rectangular tolerancing. The nongeometric (rectangular) zone must be drawn *within* the .030 circle and contain Point A.

Distance C-A is .015 inch from the center point. For a rectangular tolerance to apply, to contain Point A and all others similar to Point A, the inscribed square must be drawn. This square represents the equivalent coordinate tolerance zone—the plus and minus range of possibilities for nongeometric position.

In Unit III, you will learn to solve for the size of the square using a math tool called Pythagorean's Theorem. The square is .0212 inches across.

Conclusion—The area of the round geometric tolerance zone is 57 percent bigger than the rectangular tolerance zone. The center of a drilled hole has 57 percent more possible locations in which to land and yet the parts will assemble just fine.

The equivalent coordinate tolerance of a .030 diameter geometric position tolerance would be +/- .0106 (half of the square).

Not using geometric position tolerance for the center of the hole means that all the functionally acceptable shaded areas are eliminated. A rectangular tolerance zone simply costs more in terms of reduced tolerance.

In Figure 10-7, we see a flat feature being controlled. The tab centerline is positioned 4.5 inches basic from datum -A-. The tolerance zone is a pair of lines .010 inch apart. The centerline of the tab must fall within this tolerance zone.

In both Figures 10-5 and 10-7, how do you know that the control applies to the centerline of the feature? There are two ways. First, the obvious answer. These are controls of position, and position only deals with feature centerlines. Second, the control frame where the position is called out is adjacent to the size callout that denotes that the control applies to the centerline of the feature. We learned this drawing standard when studying controls of form. When the control applies to the centerline of a feature the callout will be adjacent to the size requirement.

Position Using MMC or LMC

In Figure 10-5 and 10-7, the position tolerance was fixed–no bonus because of the RFS requirement in the control frame. Regardless of the size of the feature, the

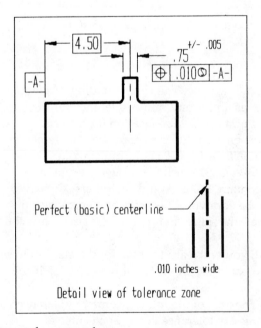

Figure 10-7 Flatness of a center plane.

position tolerance remained as shown. A bonus tolerance is frequently possible based upon the function of the part.

Only When the MMC or LMC Symbol is Present

The shop may be able to gain bonus position tolerance. This is only possible if the designer has permitted it by adding either the LMC or MMC symbol to the control. A second factor is that, only after an actual machined part is measured and the amount of deviation from either MMC or LMC size is determined, may we compute bonus tolerance. To review, the total of the bonus tolerance added to the stated position tolerance in the frame, is called Earned Tolerance.

To review the Earned Tolerance process, return to Chapter 8. The following is an outline of the procedure for calculating the full Earned Tolerance based upon an actual measured part.

If the drawing has the LMC or MMC symbol:

1. Determine the MMC or LMC size from the print. This is design size plus or minus the full tolerance. This material condition size will lie at the extreme limit of size—either the smallest or largest possible size depending on the kind of feature (internal or external) and condition (MMC or LMC).
2. *Measure the part to determine actual size deviation* from L/MMC.
3. *Calculate bonus tolerance.* This is the absolute difference in actual size and MMC/LMC size. Whichever material condition applies in the control frame. The numeric difference in 1 and 2 above. Remember this will be a subtraction of one value from the other, the numeric difference in the two values. But depending upon the material condition (MMC or LMC) and the type of feature (internal or external) they will be in mixed order. Use the absolute result.
4. *Compute Earned Tolerance.* Add bonus tolerance to stated tolerance.

Formula for Finding Earned Tolerance

Earned Tolerance = Bonus Tolerance + Stated Tolerance
Where
Bonus Tolerance = | Actual Size−MMC Size | The absolute difference

Example Problem.

Consider Figure 10–5. Suppose that the control frame was MMC rather than RFS. This signals the user that the possibility of bonus tolerance exists; that the position tolerance on the drawing is the worst case when the size of the hole is at

MMC. The .030 position tolerance is designed to fit when the hole size is at MMC. Suppose the hole is actually drilled at a diameter of (.627) inch.

What is the Earned position tolerance (stated tolerance plus bonus).
Solution:

$$\text{MMC Size} = .615 \text{ (most metal, smallest hole)}$$
$$Actual = .627 \text{ (from measuring)}$$
$$\text{Bonus} = .012 \text{ inch}$$
$$\text{Earned tolerance} = \text{stated} + \text{bonus}$$
$$= .030 + .012 = .042 \text{ inch total tolerance}$$

Solve the same problem assuming that the hole size measures .6235 inch. Use this space to do your work and label each step.

Answer: Earned Tolerance = .0385

Calculator Note: When using a calculator to solve problems of this type, it is useful to note that when finding the difference between two values, the answer will have the same absolute value (disregarding sign) no matter in which order you enter the numbers to be subtracted. You may find the difference in two numbers without regard to order of calculator entry. Try it for the above problem where MMC size was .615 inch and the actual was .6235 inch. Subtract them in both orders. The result will be either .0085 or −.0085 depending on the order in which you entered the values. Since the actual bonus will always be a positive number, if the result comes out negative, you may simply correct the minus value using the (+/−) sign change key of your calculator.
Example: −.0085 [+/−] = .0085

Inspecting Geometric Position

Determining Actual Centerline

Inspection of geometric position can be accomplished using standard methods or a CMM. The objective of the inspection is to determine if the centerline of the feature lies within the tolerance zone. The problem is that the centerline is a derived entity and in fact doesn't actually exist. It is established by the part feature size, shape, and location (shape will also determine the derived centerline position).

Deriving the centerline of the feature is mathematical and we will study this in Unit III. Inspection then requires some process that derives the centerline position of the controlled feature from the size and position of the surface of the feature. If the feature is a hole, then a set of ascending size ground gage pins may be used—inserting the largest pin that just fits into the hole, then smoothes out the irregularities and simulates the geometric hole. That is the largest round space available.

A CMM operator can find the true center of a hole by requesting the measured position in geometric terms with respect to a datum or datums.

In Figure 10-8-A and 10-8-B, we see a screen readout of the geometric position of a hole. The user makes a set-up where the CMM uses the intersection of datums -A- and -B- as the origin for dimensioning. Then, touching the probe several times inside the hole, the CMM computes the true center of the hole. The CMM can account for irregularities detected and responded with the true geometric position of the hole.

Standard methods such as a height gage can also be used. First the roundness

Figure 10-8a A CMM inspecting a hole position.

124 □ Unit II The Working Geometric Concepts

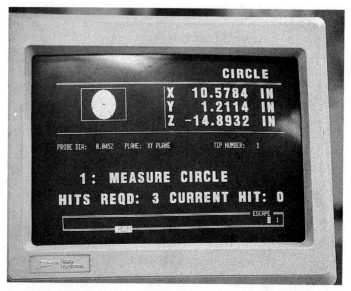

Figure 10–8b A screen readout to inspect a hole size and location (Courtesy of Brown & Sharpe Co.)

of the hole must be known, then edge measurements from datums -A- and -B- must be taken. An improved process uses a round inspection pin inserted in the hole and its center compared to the control datums. This then depends upon having the correct largest "best fit" round pin.

Position / Concentricity Challenge Problem 10–1

Answers follow.

1. In Figure 10-9, the slot is machined at .624 inches wide. What is the earned position tolerance for this part?
2. What is the earned position tolerance for the hole in Figure 10-9.
3. Figure 10-10 is a critical thinking challenge. This is a Least Material Condition function—the designer was concerned with the thickness of the support walls on this hinge bracket.
   A) Taking into account the inside slot and outside boss, what is the thinnest wall that will ever occur within design tolerance? You must account for both size and position.
   B) Assume that the slot is machined to .497 inch. What earned position tolerance applies to this part?

CHAPTER 10 Characteristics of Location □ 125

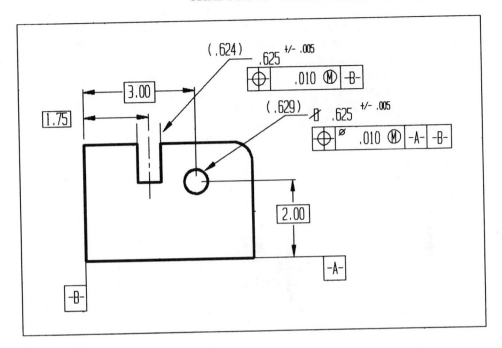

Figure 10-9

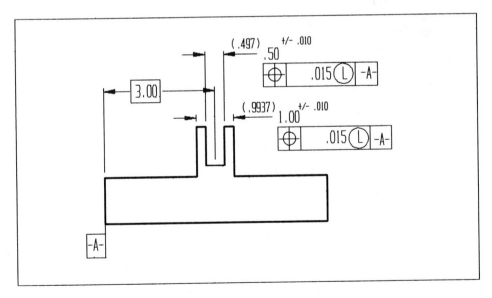

Figure 10-10 A machined part with two features.

126 □ Unit II The Working Geometric Concepts

4. On Figure 10–10, assume that the tab measures .9937. What bonus tolerance does this produce and what is the earned tolerance?
5. On the locator pin in Figure 10–11, the small diameter is measured and found to be 1.003 inches. What concentricity tolerance is earned by this feature size?
6. Suppose you now wish to allow the small diameter in Figure 10–11 to have bonus tolerance with respect to datum axis -A-? Complete the feature control frame in Figure 10–12 so the control applies MMC.
7. With the revised control frame of Question 6, re-answer Question 5 for a pin diameter of 1.003 inches.

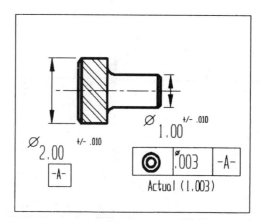

Figure 10–11 A control of concentricity.

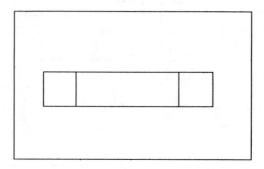

Figure 10–12 Design a control to allow bonus tolerance.

Answers to Challenge Problem 10-1

1. MMC SIZE = .620
   ACTUAL = .624
   BONUS = .004
   EARNED = STATED + BONUS
   .014 INCH = .010 + .004
2. Earned tolerance for a .629 hole is .019 inch.
3. A) First determine what the wall thickness will be with both features at LMC size, but still centered.
   LMC SLOT = .51   LMC BOSS = .99
   (.99 − .55)/2 = .24
   Now, each feature wall may vary by half the total position tolerance at LMC; ie., .0075. The wall will thin out another .015.
   This reduces the .24 by .015 = .225 MINIMUM THICKNESS
   B) The earned position tolerance for a .497 slot is
   LMC .51
   −ACT .497
   BONUS = .013   EARNED = .013 + .015 = .028 INCH
4. LMC TAB = .9900
   −ACTUAL = .9937
   BONUS = .0037   EARNED = .0037 + .015 =.0187 INCH
5. This is a control of concentricity—it applies RFS only. The concentricity tolerance remains .003 for all feature sizes. There is no bonus.
6. You need to make two changes. Add position and MMC symbols (Figure 10-13).
7. MMC = 1.010
   ACT = 1.003
   BONUS = .007 inch
   EARNED POSITION TOLERANCE = .007 + .003 = .010 inch

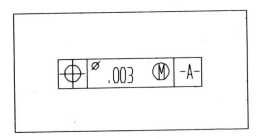

Figure 10-13 A control that allows bonus for the pin.

Review of Position

- Position is one of two controls of location.
- Position defines a perfect model location for the center of the feature.
- This perfect model position is defined from datums using basic dimensions.
- The tolerance zone is built around this perfect model.
- In three dimensions, the zone will be cylindrical for round features and a flat sandwich for centerplane features.
- The zone will be the distance between two surfaces for features such as slots and tabs where a single axis is being controlled.
- In both cases, round or flat features, the distance *across* the zone is the position tolerance.
- The centerline of the controlled feature must then fall within the tolerance zone.
- Position must apply either RFS, LMC, or MMC. None are automatic; the designer must state which condition applies with each control.
- If a material condition modifier is in effect, the position tolerance may grow based upon an actual size that is different from the LMC or MMC size.
- Inspecting position requires a mathematical process to determine the true position of the derived centerline.
- This requires measuring the size and form of the surface of the feature in question.

CHAPTER 11

Geometric System Notes

Chapter 11 contains geometric details that you will encounter in shop work. These are facts of which you should be aware as you apply geometrics in manufacturing. The ANSI Y14.5-M Chapter-Section and Paragraph will be cited for future reference or deeper study should it be required.

Only Projected Tolerance Zone and Zero Position Tolerance have been assigned concept numbers. The remaining categories are less frequently used and thus mentioned briefly.

Projected Tolerance Zone—Concept #28

Control Beyond Physical Feature

Often the influence of a feature will need to extend beyond the actual physical part. Examples of anticipated projection of influence could be a bored hole into which a pin is to be pressed or from which a bolt or other fastener protrudes. Another common feature that requires a projected tolerance zone is a threaded hole.

Consider Figure 11-1. Here we see the position of a precision hole centerline applied MMC. Suppose that this hole is to have a pin installed in it. The hole actually then becomes a cylinder whose form envelope extends beyond the actual part. The position tolerance can allow some centerline deviation from perfect angularity. Within the part thickness this is acceptable; however, as the pin is installed, the error becomes amplified. The part might not assemble.

Notice the extra control box below the position callout. This denotes the extent of the control (.75 inch) beyond the physical part: the Projected Tolerance Zone. A second method used to note a projected tolerance is to include the Ⓟ symbol callout and then to draw a dark line and dimensioning the extent of the control zone in the area affected.

A projected control anticipates the thickness of a mating part and insures correct assembly. In effect, the more critical aspect of the pin protruding from the bored hole is that portion outside of the part—the part of the pin that must mate with another part. Therefore, a projected tolerance zone originates at the actual

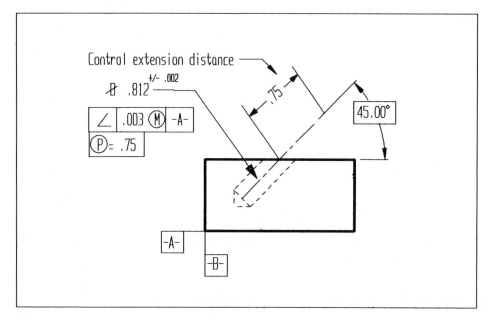

Figure 11-1 Control beyond physical part. A projected tolerance zone.

centerline location of the pin at the face of the part. This real location then becomes the model location for the tolerance zone. (See ANSI Y14.5-M 5.5.)

Inspection of Projected Tolerance Controls

The inspection of projected tolerance controls must be done using gages to simulate the projection of the feature or a CMM designed to calculate the projection. The hole example in Figure 11-1 would require the largest ground precision pin that would fit in the hole. This pin is inspected as though it was an extension of the hole. Be sure to inspect the pin no further than the noted control extension.

Zero Position Tolerance At MMC—Concept #29

Under certain conditions, the designer may choose to manipulate the dimensioning and tolerancing of a feature to allow a wider tolerance range of parts—that is, for example, to say accept parts that might be otherwise rejected because they were beyond the position (or other) tolerance yet that would assemble and function just fine. In fact, the extra possibilities are those that are actually more precision than the original tolerance.

In Figure 11-2, we have a part that must fasten to a second identical part by sliding two .250 pins through the holes. Note that the holes have clearance for the .250 diameter pins. The MMC size of the hole is .255 diameter and has a position tolerance of .005. The hole could move .005 inch in any direction. In all cases within this design, the .250 pins will fit through both parts. However, any holes less than .255 would be rejected, when in fact, if they were accurately located, they would assemble—the pins would fit.

Suppose a .254-diameter hole was positioned within .004 geometric location. Would the part still assemble? Would it accept the .250 pin and assemble to the second part. Yes, it would. In fact, a .250-diameter hole with perfect position would accept the pin (body to body) if it was exactly located on position.

Using the geometric tool of Zero Position at MMC, a wider range of parts can be accepted. Here is how this occurs.

In Figure 11-3, the dimensioning and tolerancing for Figure 11-2 has been rearranged to accommodate all the assembly possibilities. In this expanded version, the positional tolerance of the holes is zero when they are at MMC, but as they deviate they gain position tolerance. This part will assemble in exactly the same fashion as the first—with no more clearance at LMC, or at MMC—yet we have extended the possible range of parts, expanded the tolerance even though it appears that a position tolerance of zero is quite tight. The range of possibilities has been enlarged.

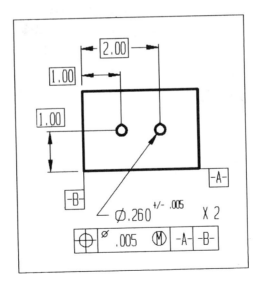

Figure 11-2 Two of these parts will be fastened together using floating fasteners.

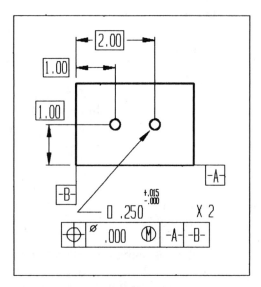

Figure 11-3 More possibilities than Figure 11-2.

See for yourself. Calculate the positional tolerance for a .255 hole in both cases.

In Figure 11-2 the positional tolerance is .005; in Figure 11-3 it is .005. The only difference is that any hole size less than .255 would be rejected in Figure 11-2, yet work fine and be accepted in Figure 11-3. Try a .257 hole. (See ANSI Y14.5-M 5.3.3.)

Composite Positional Tolerances

Consider Figure 11-4, an instrument panel that is to accept three instruments of exactly the same shape and mounting hole pattern. The cutouts must be closely toleranced within the pattern yet the position of the group of features as a whole for each cutout must not be so closely toleranced. The location of each instrument is not as important as the tolerance within the individual pattern. There are two ways to deal with this design requirement.

Reference Datum Within Pattern

Assign a central datum of size to each instrument cutout pattern. Then tolerance the pattern from this central reference. The position control tolerance of this central reference datum would be as required for the overall location of the group.

CHAPTER 11 Geometric System Notes □ 133

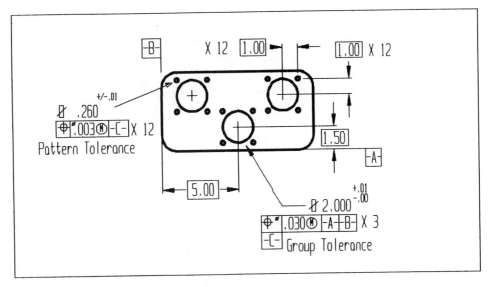

Figure 11-4 Locating feature methods.

In Figure 11-4, the group would have a pattern tolerance of .030 at MMC for datum -C-. This datum then is the control datum for the group. The mounting holes would be dimensioned and toleranced from this central datum. In effect, the small holes float with the location of the central datum. Consider the functional requirement for precise hole location. The more critical issue is the inter-pattern relationship. Each instrument mounting pattern must be closely toleranced in order for the mounting screws to fit. However, the location of the central datum is less critical—it can have more liberal position tolerancing.

As the central datum moves, it then carries the mounting holes along with it. They float with (follow) the central datum.

A Composite Positional Tolerance

A second method of dimensioning and tolerancing this instrument panel is a composite tolerance. Here a composite tolerance has been given for the group (Figure 11-5). The top control is the pattern tolerance and the bottom control of .003 is the inter-pattern control tolerance. The lower requirement would control the location of the features within the pattern while the upper tolerance of .030 inches would control the overall location of each pattern. (ANSI Y14.5 5.4.1.)

In both cases, as the mounting holes deviate away from MMC (tightest) size, they would have less accurate location, bonus position tolerance.

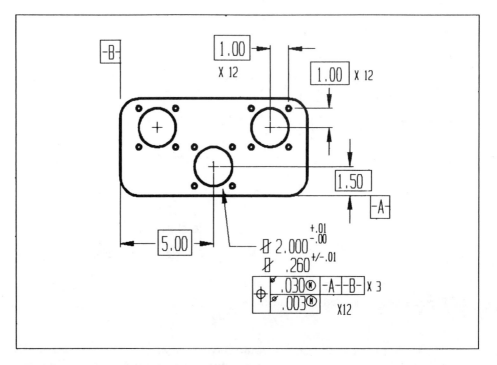

Figure 11-5 A composite tolerance method.

Conical and Bidirectional Tolerancing

Within the Geometric Dimensioning and Tolerancing System, two odd-shaped positional tolerance zones are possible. Both are caused by differential requirements for control at opposite ends of a feature. Conical and rectangular zones are the result.

Closer tolerancing at one end of a feature than another end is possible. This produces nonstandard-shaped tolerance zones; for example, a position tolerance for a reamed hole in a transmission casting. On the internal side the position is critical for assembly but on the outer end of the hole, we simply need to be within a cast boss.

A position tolerance might be used, such as that in Figure 11-6. This would then generate a conical-shaped tolerance zone. (ANSI Y14.5-M 5.8.)

Bidirectional Tolerancing

This is used when one directional component of a hole's (or any feature with a centerline/plane) location is more important than the other direction. For exam-

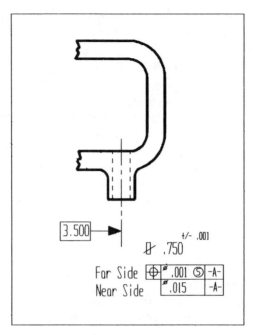

Figure 11-6 A conical tolerance zone.

ple, take a hole for a locator pin that must be accurately placed with respect to a surface but along the surface is less important. For a hole, this yields a rectangular tolerance zone rather than a circle. (ANSI Y14.5-M 5.9.)

Variation—Free State or Restrained

In calculating fits and tolerances, often the designer must analyze the function of the part or assembly. Some parts are non-rigid; in an unassembled condition, they may flex or move in some way. The example cited in ANSI is a round part with a relatively thin wall as compared to its diameter. When designing parts such as these, there are two possibilities for the method of inspection:

1. Free State or
2. Restrained

The method chosen depends upon the function of the part, whether the part is subject to control while in the assembly or when in the unrestrained condition.

For example, is the surface of a part flat enough for a watertight seal when bolted to the mating part or must it be flat with no external force?

Free State Variation Inspection

If the function of the part or assembly is such that the articles in question must be within tolerance while no external forces are applied, then the note "FREE STATE" may be drawn below the control frame. (See ANSI 6.8.)

This requires that the part or assembly not be forced into a different shape than the natural shape of the unrestrained part. In this free state, the control tolerances are evaluated. A CMM is ideal for this type of inspection because the CMM can accept the part in any orientation.

If no CMM is available, inspection is also possible but requires caution. If you use clamps, fixtures, or other holding devices, ensure that these do not change the part in any way.

Restrained State Inspection

Although the complexities of simulating the assembled condition are present, restrained inspection is simplified due to the fact that it may be held against inspection tooling or fixtures. Often, a fixture is necessary to stimulate a mating part or condition.

CHAPTER 12

A Systems Approach to Geometrics

Upon completion of Chapter 11, your building block, background education is complete. In this chapter we will review what you have learned but, far more importantly, we will look at geometric dimensioning and tolerancing from a systems perspective. Once you have the individual geometric concepts under control, it is time to see how they relate to each other; to see the entire system as a designed whole. Credit must be given to the ANSI Y14 committee for this—this subject is truly designed as a system.

Scan Figure 12-1; it is a compilation of the system. This chart will help in showing the relationships between the different characteristic control symbols. We will return to this chart several times. Viewing the entire system will deepen your knowledge and provide some intuitive feel for geometrics. We will continue this systems approach in Unit III where we will see interrelationships at work.

Note that in note 2 in Figure 12-1, there is a similarity to form controls where we observe a single and all element version of the control. If the control pertains to a centerline (hole or pin), the control deals only with the single line. If the control pertains to a centerplane (tab or slot), then the control is for the entire derived centerplane. In both cases, these controlled lines are not strictly elements because they are mathematically derived from surface elements. Elements are lines that actually exist on the surface of the part in question. However the concept is the same—two-dimensional and three-dimensional versions of the control.

Combining Controls

Perfect models must often be built using the basic controls in the system. For example, can you devise a control set that defines a perfect cone shape? Think about this challenge—what controls would you select? There are a couple of ways to geometrically dimension and tolerance a part that is a cone.

Control	Symbol	Datum Reference	Single or All element	Applies MMC or RFS	Can MMC/LMC Apply	Notes*
FORM	—	NO	SING / ALL	RFS	YES	1, 2
	▱	NO	ALL	RFS	NO	
	○	NO	SING	RFS	NO	
	⌭	NO	ALL	RFS	NO	
CONTOUR	⌒	YES/NO	SING	RFS	NO	
	⌓	YES/NO	ALL	RFS	NO	
ORIENTATION	⊥	YES	SING / ALL	RFS	YES	1, 2
	∠	YES	SING / ALL	RFS	YES	1, 2
	∥	YES	SING / ALL	RFS	YES	1, 2
LOCATION	◎	YES	CENTERLINE CENTERPLANE	RFS	NO	
	⌖	YES	CENTERLINE CENTERPLANE	MUST BE RFS MMC LMC	YES	
RUNOUT	↗	YES	SING	RFS	NO	
	↗↗	YES	ALL	RFS	NO	

Figure 12-1 The geometric control characteristics.

[1] Applies to surfaces but can also control a feature centerline or centerplane.
[2] If application to a centerline then the control applies not to an element but to the single *derived* centerline.

If applied to a centerplane, then the entire derived plane is controlled.

You are trying to control an entire surface so the set must be an ALL element control. In effect, you are designing a control very similar to cylindricity yet with tapered sides.

Could you use

— Roundness and Angularity?
— Roundness and Profile of a Surface?

Co-axial controls could work in some cases because of the need to control the roundness of the surface. For example, runout actually embodies a control of roundness as well. Although not defined in ANSI, due to the angular surface, could you use:

— Total Runout and Angularity

There are other combinations that would work as well. Remember this example; you will be asked to repeat it soon. In combining controls, the designer must exercise care to not over-control nor create a set that actually conflicts.

Example:

In Figure 12-2, there is a 5/8 inch hole that is controlled for form, position, and orientation to two surfaces.

This is an incorrect control with too many requirements. Engineers try to avoid controls such as this due to the complexity in quality control and the conflicts that it creates. We will work with this drawing again in the problems at the end of the chapter to see if we can simplify it using interrelationships and taking advantage of embodied controls.

Understanding these interrelations within the geometric system is vital to a complete working ability, especially for total quality assurance. It would be wise to read the following definition several times and know it well.

A Final Definition of Geometric Dimensioning and Tolerancing

The geometric system controls dimensioning and tolerancing of 13 design *characteristics* of manufactured parts. In all cases, a *perfect model* of the characteristic is defined. A tolerance *zone* is then built around the perfect model. Either surface *elements* or *centerlines/planes* must then fall within this control zone. Controls are either single or ALL element (single line or all lines for center feature controls). The entire system is based upon function and fit.

Based upon true functional requirements, the GDT system is designed to derive the *full* natural tolerance in the design of the assembly or detail. Further, the GDT system allows full tolerance to reach the manufacturing floor thus producing less scrap due to more possibilities of good parts. These extra possibilities come from two related aspects of GDT:

1. The tolerance range for a GDT part compared to nongeometric design is greater due to the natural tolerance reaching the shop floor and due to the bonus tolerances GDT provides.
2. The control points—the dimensions and related tolerances—are based on the function of the parts. Priorities can be custom suited to the design. This means simpler manufacturing and more control of the process.

140 ☐ Unit II The Working Geometric Concepts

Because GDT is based totally upon the functional requirement of the design. The result is cost reduction in manufacturing while making better quality parts.

The 13 Characteristics

Sorting By Similarity

In Figure 12-1 you see a compiled chart of the 13 post-1982 characteristic controls. It is interesting to see how they relate to the system in general. We will perform a sort of these characteristics to see how they relate to each other, work together, and complement each other. We will run this group of 13 characteristics through four sort categories:

1. Datum Requirements—Yes or No
2. Single/All Element Requirements
3. Automatic Material Condition Applicability (when no modifier is present)
4. Is a Bonus MMC or LMC Tolerance Applicable

Please scan Figure 12-1 at this time. You will be given some critical thinking questions to help analyze the system. From these questions, you can draw some conclusions about the geometric system for yourself.

Unit II Self Challenge Problem 12-1

Final Evaluation

Note: this material is designed to help review what you have thus far learned. If you cannot answer a problem or feel uncomfortable about your answer, then return to the chapter where the characteristic is covered. The best place to verify your answers to these problems is in the text material.

Brief answers will follow this activity. In many of the questions you will be asked to write a brief statement of what you understand.

This is a good time to appraise your success in studying this material because we will be entering Unit III next, where you will need this knowledge to continue up the critical ladder. If you can answer the majority of these questions with an accuracy of 80 percent, and you understand the answers provided at the end of this set to the other 20 percent, you are ready to proceed into the more challenging work in Unit III.

Please note that these are critical thinking questions and you may not be able

CHAPTER 12   A Systems Approach to Geometrics   □   141

to answer immediately without some thinking and re-reading. You will be learning about geometric systems as you respond—this is a mastery exercise.

1. Notice that in sort field 1, "Datum Reference," there are two controls that may or may not require a datum as a reference. Can you explain why?
2. Why do the first four controls in Figure 12-1 not require a datum reference?
3. For a control of concentricity, does the control datum project out into the controlled feature and then contain the centerline of the feature—or are we projecting the controlled centerline into the reference datum?
    (This question is not directly covered in the text material, but think about it—you can answer from what you know.)
4. In sort 2, how many controls are always *single element*? Why would you not include controls to which note 1 applies?
5. When is it correct to apply an MMC requirement to a control of straightness?
6. Why are "contour of a line" and "contour of a surface" separated from form even though they are very similar?
7. If you could choose only two controls as master controls for the first six on the chart, which two would it be and why?
8. If you could only choose one control of orientation as a master control, which would it be and why? Using only this master control, what change would this make on the drawing?
9. Why are both concentricity and position included in controls of location? That is, what similarities do they exhibit? What dissimilarities?
10. When can parallelism apply at MMC?
11. What is the difference between concentricity and runout?
12. Suppose you need to control a cone-shaped object. What combination of controls would you select to build a "Cone-Endricity" control?
13. By controlling cylindricity, what other feature characteristics are you also controlling?
14. When are controls of orientation not *ALL element*? Name two instances?
15. Is this statement True [ ] or False [ ]? If it is false, write a correcting statement.
    Controls of location must always refer to datums and control centerlines of the feature.
16. To how many controls may an MMC/LMC bonus tolerance be applied? List them from memory before checking Figure 12-1.
17. Explain note 1 for Chart 12-1.
18. This is an advanced question. How may we simplify the control for the 5/8-inch hole in Figure 12-2, yet get the same result? Draw the control on a separate sheet.

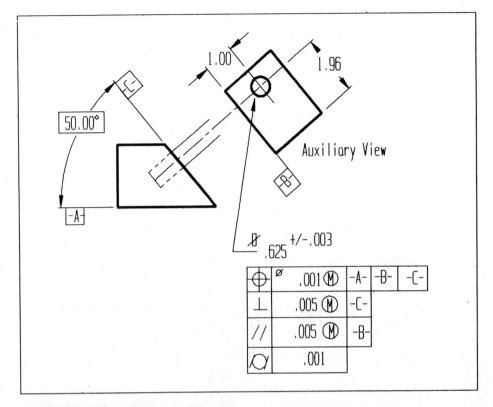

Figure 12-2 An overly complex control.

19. Is this statement True [ ] or False [ ]? If false, write a correcting statement.
    The material state modifier Least Material Condition (LMC) is mostly used to control the weight of designed products.
20. Advanced Challenge (Figure 12-3). Part A is a perfect gage whose slot is .500 wide and centerline exactly 1.5 inches above datum -A-. Your challenge is to dimension, control, and tolerance part -B- such that it will always assemble to gage A.
    The assembly has the following functional requirements. Both parts must lie on Datum Feature Surface A when assembled. Part B must always maintain a minimum clearance of .001 inch per side at MMC size and a maximum clearance per side of .005 inch. *Part B is the only part that varies. Part A is a functional gage for this exercise.*

    Although in real design, irregularities on datum surface -A- and form and orientation of the tab would also affect the design, we will eliminate

CHAPTER 12 A Systems Approach to Geometrics □ 143

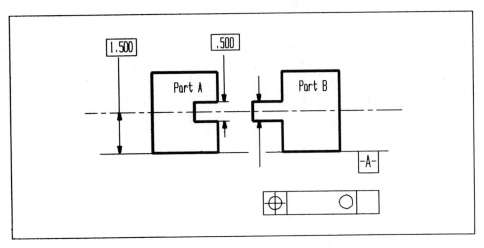

Figure 12-3 Complete the size and position.

them from the exercise. Work with only size and position. The positional requirement may or may not be MMC as you choose but your choice will affect the .005 per side LMC clearance and thus the size and position ranges.
21. In Figure 12-2, based upon nothing more than what you see, what might be the reason for the MMC application of the position control?
22. Without looking at Figure 12-1, complete the blank chart in Figure 12-4.
    — Name the control group; i.e., form, etc.
    — Sketch the geometric symbol
    — Indicate whether datum reference is needed
    — Show whether the control is single or all element
    — Check either RFS or LMC/MMC for each control
    — Indicate where MMC/LMC bonuses can apply
    (Student note—you may wish to photocopy Figure 12-4 now before you attempt to complete the blanks. You may wish to do this exercise more than once, especially if you do not get over 80 percent the first time.)
23. What two shapes may the control zone assume for straightness? Which is the more common?
24. Write your own definition of the geometric system. Your definition should include the words: *Characteristics, Perfect Model, Elements , Centerlines, Centerplanes, Functional Fit,* and *Tolerance Zone*
25. Is this statement True [ ] or False [ ]? If it is false write a new statement that will make it true.

Control	Symbol	Datum Reference	Single or All element	Applies MMC or RFS	Can MMC/LMC Apply	Notes*
FORM	—		ALL			1, 2
	⌐					
	⌒					
						1, 2
						1, 2
	//					1, 2
		CENTERLINE CENTERPLANE		RFS		
		CENTERPLANE		MUST BE RFS MMC LMC		
RUNOUT						

Figure 12-4 Fill in the blanks.

The term "Basic" means a dimension that is more important to the design function of the part than other dimensions. A basic dimension will have the tightest tolerance on the drawing.

Answers To Challenge Problem 12-1

1. These are controls of contour—Profile of a Line and Profile of a Surface. Functionally, the contour may need to be orientated or located to some other feature. This control feature, then, becomes a datum to the contour.
2. Form is a floating control. It is a comparison of a perfect counterpart of itself, thus no datum is required. In other words, straight is simply straight with no outside influence required.
3. The perfect control datum axis projects out into the controlled area of the part. The datum is perfect, thus projecting it out to the control area creates

no errors. Doing the reverse—projecting the controlled axis error to the datum—would amplify the error due to the projected extension.
4. Three controls. Roundness, Contour of a Line, and Circular Runout. The key word is *always*. Other controls are usually single element; however, they may apply to a centerplane where multiple lines are controlled.
5. When the control applies the centerline or plane of a feature of size.
6. Because they may or may not require a datum. In actual fact, contour is a form control; however, based upon function, the form may require datum referencing. That makes it a separate control.
7. Contour of a line and contour of a surface. However, then to control straightness, flatness, roundness, and cylindricity would always require a drawing definition for each. In other words, when controlling straightness, you need no definition of straightness. However, using contour to control straightness would require a drawing definition each time it was used.
8. Angularity. Similar to the previous answer. For example, using angularity to control perpendicularity would always require a definition of perpendicular each time it was used. The basic angle (90 degrees) would need to be stated each time.
9. Similarities:
    Both control the centerline or plane of a feature.
    Both require datum referencing.
   Dissimilarities:
    Concentricity is always applied RFS.
    Position may be applied MMC or LMC, thus a bonus possibility.
10. When it applies to the centerline or plane of a feature.
11. Both controls verify co-axiality. However, concentricity controls a derived centerline compared to a datum axis. Runout simply compares and controls surface elements to a center axis datum.
    Concentricity is determined through a lengthy inspection process that derives the true centerline of the controlled feature and compares this centerline to a datum axis centerline. Runout is found by rotating the part against an indicator.
    You should note that it is most common to inspect concentricity by rotating the part or indicator while observing the indicator. This is truly runout—not concentricity.
12. You could use Angularity and Roundness.
13. Straightness, Roundness, and, to some extent, Parallelism of the respective opposite side elements
14. When they pertain to the center*line* of holes or pins.
15. It is mostly true that controls refer to datums and control centerlines. One minor exception is when the control pertains to a centerplane.

16. Five controls may have a functional bonus applied. Straightness, Orientation Controls, and Position
17. The control not only applies to surface elements but can also apply to derived centers of features.
18. First, we can eliminate the perpendicularity control to datum -C-. It is superfluous due to the tighter position control to datum -C-. Note that position to a datum plane perpendicular to a feature centerline automatically implies perpendicularity. Since this position control is .001 inch and the perpendicularity control is .005 inch, it is unnecessary.

    Second, for the same reason, we might eliminate the parallel control. This, however, would require further examination of the function of the part.
19. False—LMC functionally controls wall thicknesses for strength.
20. There is no single answer to this question. However, you should be able to test your design control yourself. This is done using a test called virtual condition. First, test the dimensioning and tolerancing for MMC condition. Invent a part with largest tab allowable per your design; now position it as far from perfect as allowable. This must yield a clearance equal to .001 inch.

    Next, make an LMC analysis. The smallest (LMC) tab, perfectly centered, should not produce any more clearance than .005 inch per side.
21. Functionally, as the hole deviates from MMC, the position becomes proportionally less critical; thus a bonus position tolerance is possible.
22. See Figure 12-1. You should be able to complete nearly all the chart at this point. Review then try again if you had a problem.
23. The most common control zone shape is—*the distance between two perfect straight lines* for the straightness of a single axis. Also—The distance across a cylinder is used but is less common.
24. See the final definition in this chapter.
25. False. Although a basic dimension is always important, it may not be as critical as the question implies. It simply means that the value given is the perfect "target" value around which the tolerance is constructed.

UNIT III

Application Skills in Manufacturing

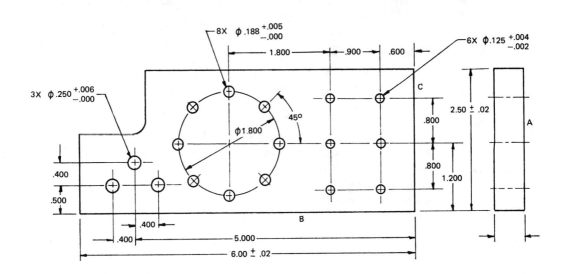

Unit Introduction

Now we switch to problem solving. Here we bring together all the basics. In this unit, we will learn the application skills of geometric calculation and manipulation. Up to this point you have been learning building block, "understanding" skills. With the exception of computing earned tolerance, you have been learning preparatory concepts.

The new "doing" skills gained in Unit III are difficult to come by without training yet they are vital if you are to apply geometrics to its fullest in manufacturing. These are the enabling tools.

We will explore real manufacturing problems and learn to evaluate the condition of a real part. This process will tell you if a part is acceptable "as made," needs rework and how to rework the part, or if the part is beyond tolerance and cannot be fixed—scrap.

This unit will teach you how to complete an inspection balance sheet for a part. In this balance sheet process you must compare the geometric tolerance available (based upon the material condition and feature size) to the tolerance used. In other words, find if the part is within geometric tolerance. You already know something about earned tolerance; now you will learn to verify geometric compliance to a control. As an example, you will find the geometric position of the feature and compare it to the earned tolerance. In this book, I refer to this process as the "Earned versus Burned" comparison.

Unit III provides real application skills in geometrics. You will learn to make decisions that can save thousands of dollars and hours of time. Can you use the geometric system to rework the part? Is the part acceptable as is? Must the part be scrapped and a new one started?

A Suggestion about Studying

This material is actually quite easy to apply. However, explanations may seem lengthy. Read through them and work the problems. The work problems are part of the instruction method and will provide critical thinking for your understanding. After successfully completing the problems, re-read the material; it will make a great deal more sense to you and will stay with you much longer in a usable way. Also, keep this book nearby as a reference as you use these skills. Review the chapter that applies when attempting to solve geometric problems.

This is the challenging, exciting stuff of geometrics. Proceed as a puzzle solver and enjoy the challenge!

CHAPTER 13

Single Axis Feature Inspection and Rework

To inspect a feature you must determine what positional error has been produced in machining a feature. That is, how much positional tolerance has been used up? Is it within the design tolerance? Then, either accept the part as being per-tolerance or determine what rework is possible.

In Chapter 13, we will look at features such as slots and tabs that have a single datum reference and a single basic dimension to denote the perfect position and determine the tolerance model.

Initially we are limiting our position study to two-dimensional problems. Therefore, the tolerance zones will be parallel lines equally spaced around a perfect centerline model. At the end of Chapter 13, we will look at some three-dimensional effects of orientation and inspection of products.

Inspection of Position

When inspecting the position of a feature, the objective is to determine where the actual centerline lies with respect to a perfect target. The distance the actual centerline falls from the target forms the geometric "Radial" error. This is the error per side from the target. The error must be less than the tolerance to be an acceptable part. In Chapter 10 we learned that the physical distance of a feature centerline from the target position must be doubled to determine the geometric error of a feature. We will designate this physical distance to one side of perfect as the Radial Error distance (RE).

The RE distance equals the physical difference in the target position and the actual machined centerline. The RE must be doubled to determine the actual geometric error produced in machining a feature.

Inspecting a Tab (Figure 13–1)

First find the RE distance by measuring the feature to determine where the real centerline actually lies with respect to the controlling datum. We need to

150 □ Unit III Application Skills in Manufacturing

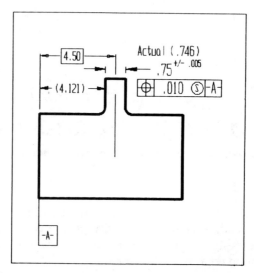

Figure 13-1 Find the radial error distance of this feature centerline.

determine the actual position of the centerline of the tab to see if it is within the .010 positional tolerance. Please solve this for yourself before reading the calculations.

We know that the distance from datum -A- to the side of the tab is 4.121 inches and the tab is .746 inches thick. Is the tab centerline within the .010 geometric positional tolerance?

Solution (see Figure 13-2)—The answer is NO. It is beyond the maximum tolerance by .002 inch.

Step 1. *From measurements, compute actual center position of the feature.* Add half the .746 measurement to the edge measurement.

$$.746/2 + 4.121 = 4.494 \text{ Actual Position} = RE = .006$$

Step 2. *Compute Geometric Error (GE = 2 × RE).* Compare the actual center to target position.

$$4.500$$
$$4.494$$
$$.006 \text{ inch RE}$$
$$2 \times .006 = .012 \text{ GE}$$

CHAPTER 13 Single Axis Feature Inspection and Rework □ 151

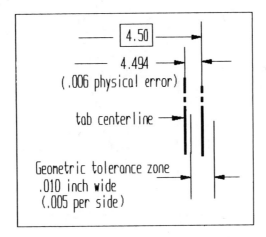

Figure 13-2 The tab centerline produces an RE distance of .006 inch which becomes a geometric error of .012 inch.

Step 3. *Compare Geometric Error to tolerance available.*

.010 Tolerance   .012 Error

The feature is beyond maximum allowable tolerance by .002 inch geometric error.

Challenge Problems 13-1

The answers immediately follow.

For the following problems, follow the three-step procedure as outlined above.

1. What is the positional geometric error for the .250 slot in Figure 13-3?
    The feature centerline is measured to be 1.7513 inches from datum -A-.
2. Considering the MMC bonus, is this part in tolerance? What is the earned tolerance available (see Figure 13-3)?

Student Note: At this point, after answering questions 1 and 2, you have completed a simple balance sheet for the slot. You computed the tolerance used (burned) and compared that to the tolerance Earned.

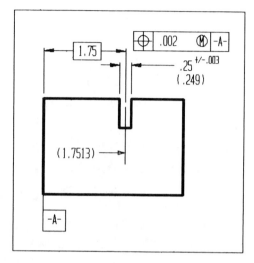

Figure 13-3 Find the geometric error.

3. What is the positional tolerance burned for the .937-inch dimension boss in Figure 13-4? (What is the geometric error?)
4. The boss in Figure 13-4 is positioned MMC.
    Is the boss within positional tolerance according to your balance sheet? That is, earned is greater or equal to burned. What is the earned tolerance? Actual size = .934 inch.
5. The left side wall of the slot in Figure 13-4 measures .313 inches. The slot measures .3118 inches. Compute a balance sheet for the slot.
6. Complete a balance sheet for Figure 13-5.
7. Now complete a balance sheet for Figure 13-5 but with MMC applied instead of RFS.
8. Recall that MMC/LMC could apply to controls other than position if the control is applied to the centerline of a feature. What were these controls?
9. Compute a balance sheet for Figure 13-6.
    The feature is inspected by the method below and found to be out of square by .0043 inch.
    A .622-diameter test pin is inserted in the hole while holding the block against datum feature -A-. Next an indicator is swept over two points along the pin. The points must be the same distance apart on the pin as the depth of the hole. The TIR difference represents the perpendicularity of the hole axis. This assumes a gage quality test pin.

CHAPTER 13 Single Axis Feature Inspection and Rework □ 153

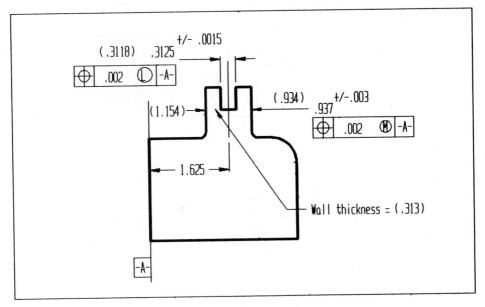

Figure 13-4 Compute a balance sheet for this part.

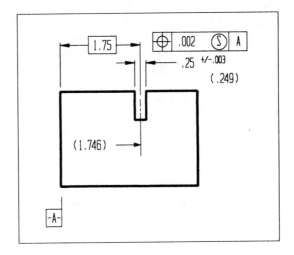

Figure 13-5 Compute a balance sheet for this part.

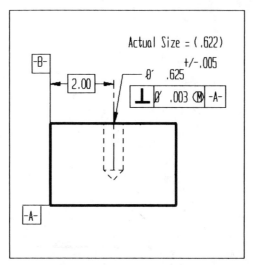

Figure 13-6 Compute a balance sheet for this part.

Answers to Challenge Problem 13-1

1. Positional error = .0026 inch (Figure 13-3)
   It is beyond the *print* allowance at MMC but not necessarily out of tolerance due to the bonus possibility.
2. Earned Tolerance = .002 + .002 = .004 inch
   Given + Bonus = Earned
   The feature in Figure 13-3 is in tolerance as machined.
3. .004 Radial Error x 2 = .008 geometric error (Figure 13-4)
   (1.154" + .934/2" = 1.621" Actual position–.004 RE)
4. YES. Earned Tolerance = .002 + .006 = .008 (Figure 13-4)
   Burned Tolerance = .008
   This feature is just within tolerance.
5. Burned Tolerance = .0042 inches (Figure 13-4)
   Earned Tolerance = .0042
   This feature is in tolerance.
   Did you use the *LMC* size to compute earned tolerance?
   Burned Calculations
   1.154 + .313 + .3118/2 = 1.6229 actual position
                                  1.625 perfect position
       Absolute Difference   .0021 RE
                                        .0042 GE

Earned Calculations
LMC is largest slot   LMC size= .3140
                       Actual size= .3118
                       Bonus= .0022
Earned = Bonus + Stated Tolerance
.0022 + .002 = .0042

Organize and lay out your work as above. This process will get more complicated as we progress, and it is necessary to label and organize in order to keep things straight in your mind. Further in the text, there will be a form to guide you through the process.

6. Figure 13-5
Burned = .008
Earned = .002 (out of tolerance .006 inch)
Note: the control was RFS—there was no bonus!

7. Figure 13-5 (with MMC applied)
Burned = .008
Earned = .004 (out of tolerance .004 inch)
MMC produced a bonus of .002 inch but that wasn't enough bonus to save the part. Later we will concentrate on the calculations to determine re-work for parts such as this.

8. Straightness, Parallelism, Angularity, and Perpendicularity.

9. Figure 13-6
Burned = .0043
Earned = .005 (Feature is in tolerance)

Form and Orientation Must Be Considered

There is a complication involved, that must be dealt with during actual shop inspection, that we have avoided for early learning examples. The inspection of the perpendicularity of the hole in Problem 9 in Challenge Problem 13-1 illustrates this concept. The orientation of the test pin was a composite of several factors. The final orientation of the pin was the composite result of the orientation of the cylindrical hole and its irregularities. The pin represented the "geometric hole."

To determine the derived centerline of a slot, for example, the sides of the slot must be treated as implied datums because they have irregularities (Figure 13-7). The derived centerline is functionally determined by the composite perfect space available; in other words, the smallest space that would accommodate a perfect fit mating part.

This "functional" slot would have parallel perfect sides. Actually the sides would be the implied datums established by the feature irregularities on each side of the slot. In Figure 13-7, we see an exaggerated slot with irregularities. The functional centerline A–B is centered between lines C–D and E–F. Line A–B would

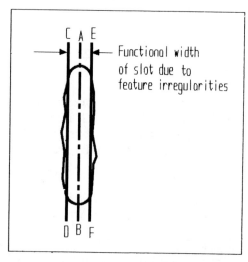

Figure 13–7 Many factors produce the *geometric* feature

be the functional center of the largest parallel block that could be inserted into the slot. From this example, you can see that functionally the sides of the slot, or any feature you wish to inspect, must be established similar to implied datums. This can be dealt with when using standard methods as long as you remain conscious of the goal—establishing the geometric position and orientation of a feature.

A CMM deals with these feature irregularities when it calculates the true position or orientation of a feature. It uses a mathematical tool called an algorithm (a mathematical analysis method) to fix the functional centerline of a feature, accounting for the feature's irregularities. If you do not have a CMM, there are ways of dealing with feature irregularities that require you to treat the feature as an implied datum. You may also find that in some cases, the form and orientation deviations can be insignificant enough for small features that you may proceed as though the establishing surfaces were perfect.

Reworking Single Axis Features

Two Ways to Rework "Out-of-Balance" Features

Once the balance sheet is completed, and an out of tolerance condition is detected such as in problem Number 5 above, the next step is to determine if rework—remachining the feature to bring the positional error within tolerance—is possible.

CHAPTER 13 Single Axis Feature Inspection and Rework □ 157

Using a nongeometric design, there is only one way to rework the feature—machine enough material to reduce the error within tolerance.

However, using a geometric design the balance may be restored in two ways:

1. *Reduce* the Burned geometric tolerance
   Machine material off one side to reposition the feature.
2. *Increase* the Earned geometric tolerance
   This is due to the increased bonus due to deviation from MMC if it is in effect.

While re-machining the feature, the amount of permissible tolerance grows while the error shrinks. Of course, this depends upon the control being MMC. If the control is RFS, only method one—error reduction—is possible. If the control is MMC, then you will need to machine less material to correct the error because of the increase in bonus tolerance gained due to the rework machining.

Any rework of an MMC feature will deviate away from MMC size, thus a gain in bonus will occur. You will soon see that LMC rework does not work that way. To establish balance on an out-of-balance LMC feature is impossible.

Reworking Features Where No Bonus Tolerance Applies

Establishing a Balance Between Burned and Earned

The objective is to bring the balance sheet into balance—that is, to make the earned tolerance equal to or greater than the burned. The entire amount beyond balance must be made up by removing material. The control is RFS—no bonus will be gained in the rework. You must compute the least amount of necessary re-machining. You must also determine on which side of the feature rework must be done.

Moving the Centerline to Reduce Position Error

Please review Figure 13-5 and problem numbers 6 and 7 in Challenge Problems 13-1 as we will use them to illustrate the difference in rework computations between RFS and MMC controls.

Rule of Sides

We have .008-inch geometric error and only .002-inch tolerance in Figure 13-8. We are beyond balance by .006 inch. To save this part we must remachine. How much?

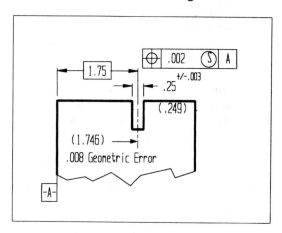

Figure 13-8 The physical facts earn .002 inch for the centerline position tolerance. From the problem of Figure 13-6.

You must understand that the amount machined off the feature changes the geometric error by the same amount. If .006 inch is machined from the correct side of the slot, the geometric error is reduced by .006 inch.

Machining .006 from one side actually moves the real slot centerline .003 inch, but recall from the previous examples that moving the real centerline toward the perfect centerline closes the zone—that is, reduces the Geometric Error by twice Radial Error distance. In effect, the geometric error zone is shrinking from both sides toward perfect as the centerline approaches perfect.

The amount machined off one side of a feature reduces the geometric positional error by an equal amount. This we will designate as the "rule of sides."

Example Problem, Figure 13-9.

This is an exploded view of the problem where we were out of tolerance .006 inch. The earned was .002 RFS and the burned was .008 inch.

You need to move the .249-inch slot centerline .003 inch to the right on the page just to be within tolerance. (In practice you would probably go just a bit more.) This is accomplished by machining .006 inch off the right side of the slot. Widen the slot by .006 inch—reduce the geometric error by .006 inch.

Question: Can you widen the slot by machining off the right side by .006 inch (shaded area)?

CHAPTER 13 Single Axis Feature Inspection and Rework □ 159

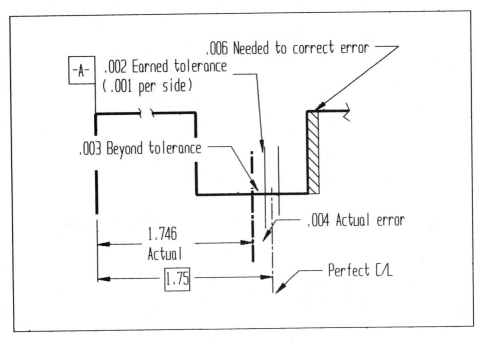

Figure 13-9 Exploded view of error.

Answer: The .006 inch off the right side would indeed bring the centerline within position tolerance but it would widen the slot to .255 inch. The upper max for this slot was .253. This part is not repairable—it is scrap!

The amount we need to physically move the centerline is shown as the distance between the actual centerline and the tolerance limit (.003 inch on one side.) To move this centerline .003 inch requires .006 inch taken off the right side. This will reduce the error by .006 inch.

Suppose a tab centerline is inspected to be geometrically out of position tolerance by .020 inch. Then machining .020 off the correct side would reduce the error to zero. If we machine .015 from one side the error would be reduced to .015 inch. Machining .010 from one side would result in .010 reduction, and so on. Your objective is to determine the minimum amount to take off the feature to bring the burned error within the tolerance earned.

Check Size Range Before Machining

To machine the .006 in the example above, we first had to check the size and related size tolerance to see if it was possible.

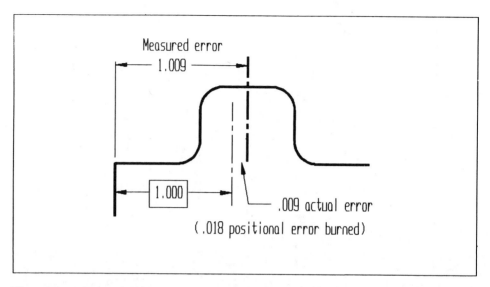

Figure 13-10 How much material must be removed to balance an earned tolerance of .010" RFS? From which side?

Example Problem, Figure 13-10

If you determined that the geometric error for a boss was .018 inch beyond perfect (.009 error to one side of perfect), from which side and how much of the boss would you machine to reduce the error to balance an earned tolerance of .010 inch RFS?

You need to reduce the geometric error by .008 inch to balance the earned tolerance of .010 inch. You must then machine .008 inch from the right side of the boss. That will close the geometric error by .008 inch. The error will just balance the earned tolerance.

Challenge Problems 13-2

This problem set will validate the rule of sides.
The answers follow:
Remember, these are RFS controls.

1. Figure 13-11 is not in tolerance.
    Compute a balance sheet to find by how much; then find how much material must be taken from which side to rework this error.
2. In Figure 13-12, the locator tab is inspected and found to be out of posi-

CHAPTER 13 Single Axis Feature Inspection and Rework □ 161

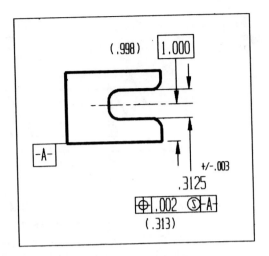

Figure 13-11 Compute a balance and rework for this part.

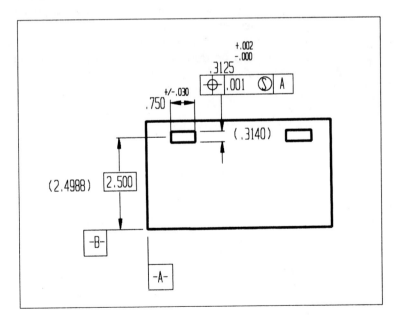

Figure 13-12 Can the locator tab on this mill fixture be reworked?

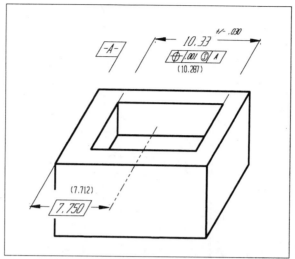

Figure 13-13 Can this pocket be repositioned exactly on center?

tion. Can it be re-machined to be within tolerance for size and position? If yes—how much and from which side?

3. In Figure 13-13, the square pocket in the injection mold base is machined off to the left. It is critical to the function that the pocket be very closely positioned but less critical as to its size.

How much material must be removed to bring the pocket *exactly* on position? Can this be done without making the pocket too big for the design tolerance?

Answers to Challenge Problem 13-2

1. <u>Figure 13-11</u>
   Geometric Positional Error = .004
   RFS Positional Earned Tolerance = .002
   Beyond tolerance by .002 inch.
   That is the amount to be taken off the top of the slot.
   .002 inch machined off the top is required.
   This is possible.
   The reworked slot will be .315 inch which is within tolerance.

   $$.213 + .003 = .315$$

   Maximum size of slot is .3155
   The reworked slot will be .0005 inch under max.

2. Yes, it can.
   Figure 13-12
   .0024 GE burned
   .0010 Tolerance earned RFS
   .0014 must be machined from the .314-inch locator.
   The reworked locator will be .3126 inch.
3. The pocket must be re-machined an additional .076 inch to bring it exactly on position (Figure 13-13). This would produce a pocket of 10.363 inches which is .003 inch too big. This part is scrap! There is no way at all to bring it within tolerance.

Reworking Features Where Bonus Tolerances Apply

Here, the feature is beyond balance and either MMC or LMC apply, thus there is a bonus tolerance in effect. This can be good news or bad news depending on the type of control—MMC or LMC. Either the bonus will get bigger or less depending on the control. In Figure 13-14, in each case, we need to machine the shaded area to move the centerline toward balance.

Answer for yourself: in the following, will the bonus get better (more) or worse (less) if we machine the shaded area from each feature?

1. On Block A, if the control is MMC, does the bonus increase or decrease?
2. What if Block A is an LMC control?

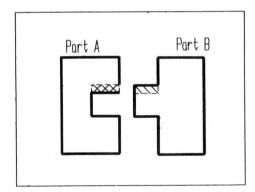

Figure 13-14 Examine what happens as you take the indicated rework cuts off each feature. See what effect each cut has on an MMC then an LMC positioned detail.

3. On Block B, if the control is MMC, is this good or bad news?
4. On Block B, what if the control is LMC?

Answers:

1. Better, because we are machining away from MMC size. We gain bonus tolerance.
2. Worse, because we are machining toward the LMC size. We are losing bonus.
3. Better, because we are machining away from MMC size.
4. Worse.

You can now see that reworking an LMC feature is impossible. With an MMC feature, as we change the size the bonus tolerance grows; but the opposite happens with an LMC feature. Here, as we remachine the feature, the bonus diminishes in direct proportion to the amount taken.

MMC Rule

When reworking MMC features—machining for example—.010 inch off one side reduces error by .010 and also increases bonus tolerance by .010. A .010 change in feature size causes .020 difference on the balance sheet.

The amount of MMC balance change from re-machining MMC features is double the amount taken off one side of the feature.

LMC Rule

When reworking LMC features, machining .010 off one side reduces the burned error by .010 but also reduces the bonus tolerance by .010. Machining .010 off one side of a feature results in *no* change in the balance sheet!

That is really bad news if, after computing the bonus tolerance and adding it to the stated tolerance, an LMC feature is out of balance (tolerance). There is no rework machining that can be done—it is scrap. What you do to move the centerline will cancel the same amount of bonus.

The only method of reworking the LMC blocks in Figure 13-14 is to plate, weld, or metal spray more material on the opposite side of the shaded areas of either Block A or B. This would then move the centerline in the correct direction while gaining bonus position tolerance. When *adding* material to an LMC positional control feature, you could apply more bonus because you would be deviating away from LMC.

Calculating Rework of MMC Features

From the example of Figure 13–14, you see that the bonus tolerance increases as you re-machine an MMC feature. This fact lessens the amount required for the re-machining of features that are beyond positional tolerance. In fact, it can reduce the amount by one half.

MMC Rework Example, Figure 13–15.

You have already solved this feature for a balance sheet. The feature is beyond balance by .004 inch (amount due). A .004-inch reduction is needed to bring the burned versus earned for the feature just within the positional tolerance.

To compute the required depth of the rework cut, note that the position control is MMC. There will be a gain in bonus tolerance as more material is removed. By machining one side only, the error is reduced while the tolerance is increased.

Divide the amount due in half. The required depth of cut is .002 inch. In practice, if possible, the cut would be just a bit more to ensure a correct position. Remember, before machining you must check the size of the feature to be sure that the correction amount you need to remachine will not then put the feature size beyond tolerance, thus producing a scrap part with no possibility of rework.

In this case, after the rework, the slot would be .251 inch which is well within the size tolerance. To check your understanding, re-compute the balance sheet. Calculate for the new .251-inch slot with a position of 1.747 inch. (The centerline radial change was .001 inch for a cut of .002 on the side of the slot.)

The burned equals .006 inch and the earned equals .006 inch.

To review the process for calculating amount of rework material for MMC features:

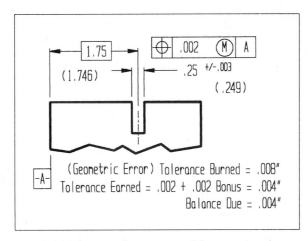

Figure 13–15 For an MMC feature, the amount of feature size change may be one-half the amount due.

1. Compute burned tolerance
2. Compute earned tolerance
3. Compute balance
    If burned exceeds earned, the amount beyond balance must be corrected. This is the balance due amount.
4. Divide balance due by 2.
    This is the feature size change amount.
5. Check the actual feature size to be sure that re-machining this amount will not take it beyond tolerance for size.
6. Determine from which side to remove material.

MMC Rework Amount Formula

The MMC rework amount formula is as follows:

$$\frac{B-E}{2}$$

where B = Burned Tolerance and E = Earned Tolerance

Challenge Problems 13-3

Using the process above:

A) Determine if rework must be done to bring the feature within geometric tolerance. If rework is indicated, just bring the feature to balance condition. In real practice, you would machine just slightly more, but that is situational based upon actual conditions.
B) Calculate the feature size change amount and from which side?
C) Can the feature withstand the re-machining? If not, the part is beyond rework.

1. In Figure 13-16, determine if rework is needed.
2. Compute the balance sheet for the slot in Figure 13-17. Is the part acceptable or is rework required? If so, how much from which side?
3. Is rework indicated for the boss on the switch post bracket in Figure 13-18? If so, how much from where?
    This is a single axis control pertaining to the boss only.
4. Determine rework for the locator slide in Figure 13-19.

CHAPTER 13 Single Axis Feature Inspection and Rework □ 167

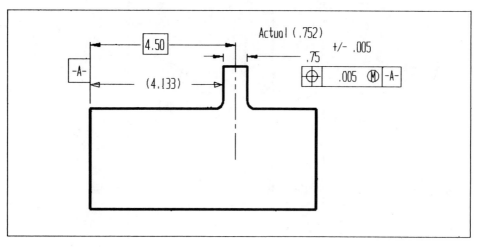

Figure 13–16 Is rework required?

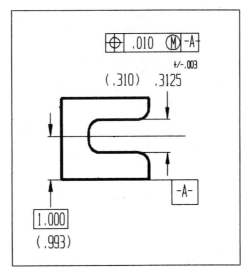

Figure 13–17 Is rework required?

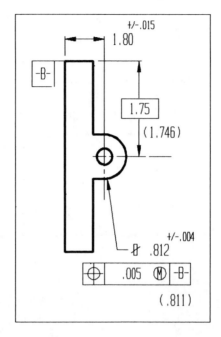

Figure 13-18 Switch post bracket.

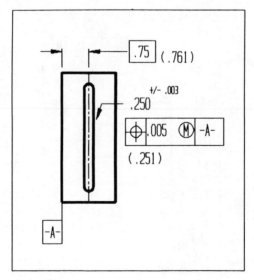

Figure 13-19 Locator slide.

Answers To Challenge Problem 13-3

1. Figure 13-16
   Burned = .018 inch
   Earned = .008 (.003 bonus + .005 stated tolerance)
         .010 Balance Due
   Feature Size Change amount = .005 inch
       Take .005 inch off right side of tab. This will reduce error by .005 and increase tolerance by .005 inch, thus correcting .010 of the out of tolerance. The reworked tab will be .747 inch which is within tolerance for size.

2. Figure 13-17
   Burned = .014
   Earned = .0105 (.0005 bonus + .010 stated)
         .0035 Balance Due
   Remachine top of groove by .0018.
   New groove will be .3118.

3. Figure 13-18
   Burned = .008
   Earned = .010
       Part is acceptable as machined. Congratulations, you finally made a good part!

4. Figure 13-19
   Burned = .022 inch
   Earned = .009 (.004 bonus + .005)
         .013 Balance Due
         .0065 Feature size change amount from the left side of the slot but we can't do that—it would make the slot oversized! This part is scrap.

LMC Method to See if Rework is Possible

A quick process to determine if an MMC feature is beyond rework is as follows:

1. Determine how much more material may be machined off the feature.
   This is the difference in actual and LMC size. This number is the total bonus that you can gain and it will be equal to the amount of reduced error.
2. Double the amount possible to machine off.
   This is the best you can do to restore the balance.
3. Compare this to the total amount out of balance.
   If the restoring amount isn't equal to or greater than amount due, then the part is scrap. You can't generate enough change to cover the balance due amount.

LMC Test Formula

The LMC test formula is as follows:

$$|A\text{-LMC}| \times 2 = \text{MAXIMUM CORRECTION POSSIBLE}$$

(A is actual size)

Max Correction Example:

For Problem 4, Figure 13-19, The Locator Slide, Challenge Problem 13-3

LMC size = .253 inch
Actual = .251
.002 inch total may be removed

Doubled, it equals the possible balance reduction .004 inch.
The out of balance amount is .013 inch. The .004 is not enough to correct the error. The part is scrap!

Review of Inspection and Rework of Single Axis Features

To compute a balance sheet and possible rework, follow this five step process.

1. Find geometric position tolerance burned—actual error.
2. Determine the earned positional tolerance. The earned must be greater than or equal to the burned tolerance. When it is not greater or equal, then rework is indicated.
3. Determine the balance due. You might also use the LMC test method at this point.
4. Determine the feature size change amount. Is the control RFS or MMC? This makes a difference. The FSC amount is the balance due for an RFS control and equals one half the balance due for an MMC control.

There are two ways to rework "out-of-balance" features:

1. Reduce the burned geometric tolerance.
2. Increase the earned geometric tolerance.
   To calculate rework amount for an out-of-balance feature:
   — If the position control is applied RFS
     You may only reduce geometric position error.

Use the direct application of the rule of sides for single axis features.
— If the position control is applied MMC

You may reduce error while increasing the bonus. In most cases this means that you need only take half the amount indicated by the rule of sides. The bonus grows as the error shrinks (is reduced). See "Special Note" below.
— If the position control is applied LMC

There is no solution. As we re-machine the feature, the bonus shrinks as the error reduces at the same rate. No change can be realized by re-machining. Plating, welding, or metal spraying are typical options for LMC rework.

5. Determine from the present size of the feature if the amount you wish to re-machine is possible. Will it take the feature out of size tolerance? You may use the quick LMC method to determine if the part is scrap. Also, you must decide from which side the material must be removed to correct the error.

Special Note on Error Correction

There are special situations whereby correcting a feature's positional error affects a second feature. The second feature is referenced from the first—the first is a datum.

We find that repositioning the centerline of the datum feature then causes the second feature to be out of position when originally it was not. In this case, we may need to machine material off both sides of the feature in varying amounts. This then gains bonus tolerance but controls the repositioning of the centerline.

In this type of rework, it is a matter of dividing the balance due amount into unequal amounts rather than dividing it in half. However, the two pieces (error reduction and bonus gain) must equal the balance due if the feature is to be within position tolerance.

CHAPTER 14

Inspecting and Reworking Two Axis Features

We will now study features that have two datums and two basic dimensions locating the perfect position. The perfect position model is a point established by the two dimension lines. We will limit our discussions to two-dimensional work; the tolerance zone, then, is a circle. Calculating a balance sheet and related rework within the round tolerance zone introduces two new concepts and one new option for rework.

These new skills to be learned are related to the fact that we are not working along a single axis for rework, but additionally must know in which direction from perfect the error occurs.

The new rework option is reaming MMC positioned holes bigger—taking material from both sides. This improves the bonus tolerance without changing the positional error.

Converting from Rectangular to Geometric Position

First Skill: Computing the Radial Error Distance

When dealing with round tolerance zones, positional error created during machining can be in any direction from the perfect location. We must deal with circular target zones and circular error. This means that you must learn to convert rectangular inspection data into geometric. When inspecting with standard measuring tools we get results such as in Figure 14-1. The errors show as an X direction and a Y direction distance from perfect. That is how far the real hole centerpoint is from the perfect position in rectangular data. The X axis error is .003 inch and the Y axis error is .004 inch. That data must be converted to a radial error.

The RE is doubled to calculate the geometric error which will be in the

CHAPTER 14 Inspecting and Reworking Two Axis Features □ 173

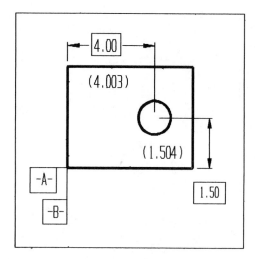

Figure 14-1 X-Y measured position shows .003 inch X error and .004 inch in Y.

form of a circle around the perfect location. The radius of this error circle is derived in one of two ways. The first way is to mathematically calculate the radius of error from XY data (see the dotted line in Figure 14-2). The second method of obtaining the geometric error of a feature is a CMM that reads directly in geometric position.

Second Skill—Direction of Error

We also need to know the direction of the error from the perfect location in order to reverse the trend upon reworking the feature. We must determine the direction in which to move the feature to recover the maximum amount of error.

Both of the skills introduced above are actually a conversion of rectangular coordinates into polar coordinates or polar to rectangular. Often polar coordinates are referred to as geometric coordinates. In this chapter we will learn to do this conversion using a calculator. It is important to learn how to convert data for four reasons:

— When checking your work in progress on the machine
— When doing rework on standard X-Y axis machinery
— For parts too large for the CMM
— To use when a CMM is not available

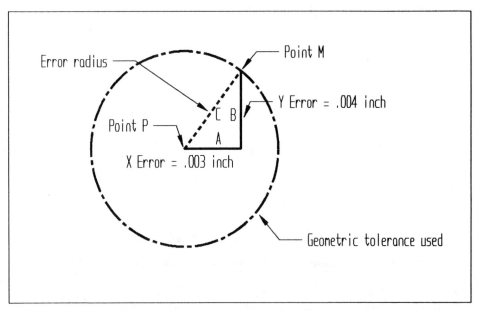

Figure 14–2 Detail view of geometric positional error. Produced by **X and Y measured errors.**

Line of Error

In Figure 14–2, when connecting a line from the perfect center point P to the machined centerpoint M, we get line P-M. This is the direction in which the positional error was generated away from perfect. To correct the error, we must reverse the machined centerpoint inward along this line of error. Rework will no longer be a simple one side or the other issue as in the previous chapter, but rather in any direction and this must be calculated.

Reaming an Extra Rework Possibility

A big advantage with holes positioned MMC is that often we may simply ream out the hole without changing the location. The larger hole increases the bonus tolerance and thus the earned tolerance covers the positional error. Reaming is an easy fix that we may select at Step 4 when determining the rework amount needed. Reaming is quick and relatively foolproof, while machining by boring to one side of the hole to reposition the center is more difficult and time-consuming. Reaming increases positional tolerance without reducing error.

Computing a Balance Sheet For a Hole or Other Round Feature—Steps 1, 2, and 3

Step 1. Finding The Geometric Positional Error

In Figure 14-1, we see an inspection record of a hole. Notice that the hole is over to the right by .003 inch and up on the page by .004 inch. Figure 14-2 is an enlarged detail view of the perfect point P and the machined hole centerpoint M. The circle represents the tolerance burned for this hole. The dotted line is the radius distance of the error. We must calculate this radius knowing the rectangular errors of .003 by .004 inch. Once calculated, the radius is then doubled to get the geometric error, the tolerance burned. The calculated circle diameter must be less than or equal to the earned tolerance to be acceptable as machined. In other words, the tolerance zone must be bigger than the calculated error circle if the part is to be acceptable. We need to convert the X and Y errors in Figure 14-2 into geometric.

Student Notes: If you know how to use a calculator and Pythagorean's Theorem to make this conversion, you might skip the next section and go directly to the Challenge Problems 14-1 to test your ability. Note, many scientific calculators also have an R-P function (Rectangular to Polar Conversion). This is a second way to make this geometric conversion. If you have this useful ability, read how to use it in your calculator manual. We will learn to convert using the square and square root functions only. But don't forget, after finding line P-M in the example, you must double it to get the tolerance diameter, the burned positional error. As an advanced challenge, if you have a programmable calculator, experiment with a program to input the X and Y factors and it converts to the diameter of the error circle either using Pythagorean's Theorem or the RP function.

Conversion Charts Not Recommended

There are commercial charts and slide rules that also make this conversion but they have limited usefulness. I cannot recommend them for anything other than as a backup to your calculator or for checking your work. First, because of the density of data, they are difficult to see and thus prone to error. Second, if you work at it, you will be much faster on your calculator and far more accurate, especially if it is programmable.

These charts require visual estimation of the answer with any conversion closer than .0005 inch. Often this isn't accurate enough. Third, these charts have a limited range of possibilities while a calculator has an unlimited range.

With just a little practice, a calculator is universal, quick, and absolutely accurate.

Pythagorean's Theorem

Figure 14-2, Finding Distance P-M

We will take advantage of Pythagorean's Theorem where, for all right triangles, the square of the two smaller sides is equal to the square of the larger hypotenuse. A right triangle is one with a 90 degree angle. For Figure 14-2, the following formula is a truth.

$$A^2 + B^2 = C^2$$

If we square Side A (multiply it times itself) and square Side B, then add these two squares together, the result will equal the square of the hypotenuse Side C. This basic theorem must be rearranged to be of use in finding the hypotenuse (radius of error) as follows:

$$C = \sqrt{A^2 + B^2}$$

Once the hypotenuse is found, we then double it to derive the positional error of the hole in Figure 14-1. This hypotenuse is the distance P-M in Figure 14-2.

Now let's see how this is done on the calculator. Ask your instructor for help if this brief discussion isn't enough, but first try the following example.

As long as your calculator uses algebraic logic, you may follow the example as shown. You can tell if your calculator uses algebraic logic if you enter a multiplication problem such as 4 × 3 = 12 in exactly the order you just read it. Nearly all calculators in student use use algebraic logic.

For Figure 14-2.

To Find the Geometric Radius of Error Hypotenuse—Side C (line P-M)

CHAPTER 14 Inspecting and Reworking Two Axis Features □ 177

Operation	Calculator Button/s	Screen Result
Enter Side A	.003	.003
Square Side A	x^2	.000009
Add Result	+	.000009
Enter Side B	.004	.004
Square Side B	x^2	.000016
Sum Squares	=	.000025
Square Root of Sum	$\sqrt{x}$	.005 Error *Radius*
Multiply Radius	X	.005
Double Result	2	
Final Total	=	Geometric Error .010

To review, we entered side A then squared it; we then touched the "plus" key to prepare to add this to the square of side B. We then entered side B and squared it. Touching the equal button sums the two squares. Now taking the square root of this total results in side C, the hypotenuse. Also note, in the future, you may enter either non-hypotenuse side A or B in any order. This formula works either way as long as you use the two smaller sides, the X and Y errors, in any order. Remember, the X and Y errors are the result of measuring a machined part and comparing the measurements to the basic dimensions.

Once the geometric error is determined, finding the earned tolerance and the balance (steps 2 and 3) are exactly the same as in Chapter 13. You will have a chance to apply them in the critical thinking problems in Challenge Problem 14-1.

Challenge Problems 14–1

Answers follow.

Note: when the problem requires you to tell where the actual positional error is located on the part with respect to the perfect position, I will call out error directions in standard quadrant nomenclature. That is, if the error is to the right and up on the page relative to the perfect position, it is Quadrant 1. Proceeding counter-clockwise, left and up is Quadrant 2; left and down is Quadrant 3, and right and down is Quadrant 4.

---

1. Solve for the geometric error using Pythagorean's Theorem.

	X Error	Y Error	Error Radius	Geometric Error
A)	.006	.0075		
B)	.0023	.004		
C)	.008	.012		
D)	.0033	.0333		
E)	.0123	.009		
F)	.005	.0000		

2. Find the geometric error for Figure 14–3.
3. Compute a balance sheet for Figure 14–4.

Note: it is a good idea to do an organizational chart such as the one shown in the answer for this problem. There is so much data that, without one, we often use the wrong numbers for calculation. As we further study rework, you will be provided a chart to keep the facts organized.

---

4. In which direction is the error for Figure 14–4? Which quadrant?
5. Without looking back, can you list the five-step process for inspecting and reworking a hole?
6. Complete a balance sheet for the hinge bracket boss in Figure 14–5.

Note: (of interest only—will not affect problem): By adding the third datum -C- in the position control, the engineer is applying a three-dimensional control. The tolerance control is extended to datum surface -C-. This then implies perpendicularity to datum -C- and a cylindrical tolerance zone for the length of the hole.

CHAPTER 14 Inspecting and Reworking Two Axis Features □ 179

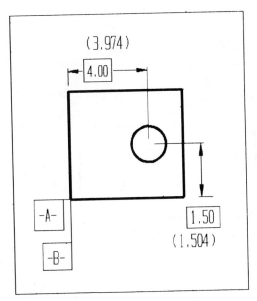

Figure 14-3 Find the geometric error.

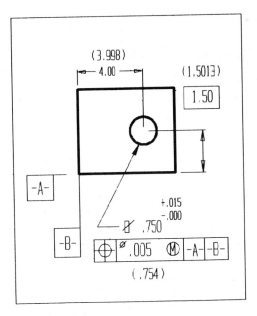

Figure 14-4 Compute a balance sheet.

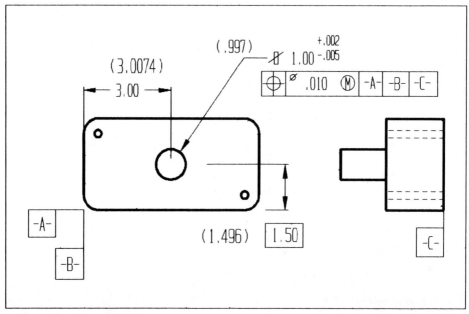

Figure 14–5 Hinge bracket boss.

Answers to Challenge Problem 14–1

(Rounded to the nearest .0001 inch. Recommended for machining work.)

1.   Radius   Geometric Error
   A) .0096 × 2 = .0192 inch
   B) .0046 =   .0092
   C) .0144 =   .0288
   D) .0335 =   .0669
   E) .0152 =   .0305
   F) .005 =   .010

   Note: although Pythagorean's Theorem worked on this problem, the error was along a single axis and did not require calculation other than doubling the error.

2. A common mistake made is to use the inspection numbers as errors but they are not. The X and Y errors are the absolute difference between the basic dimension and the inspected position.
   Example: $|4.000-3.974| = .026$ (X error)
                 $|1.500-1.504| = .004$ Y error
   Geometric Error = .0526 diameter.

CHAPTER 14 Inspecting and Reworking Two Axis Features   □   181

3. X error = .002   Y error = .0013
   Radius = .0024   Geometric Error = .0048
   Positional tolerance burned = .0048
   Positional tolerance earned = .009 (.005 +.004 bonus)
   This part is in tolerance.
4. The error is up and left from the target—Quadrant 2.
5. *First Step:* Find radial error from inspection and calculation
   Second Step: Find the geometric error—tolerance burned (2 × RE)
   Third Step: Calculate tolerance earned
   IF: Earned is greater or equal to burned the part is acceptable.
   Earned is less than burned—consider rework.
   Fourth Step: Determine balance amount due to correct out of balance condition. Is the control RFS or MMC?
   Fifth Step: Determine if the size of the feature will allow this rework without going beyond size tolerance.
   From this point on in the process, the feature centerline must be moved. You must know in which direction from perfect. Which quadrant? How much?
   Note: it is more important that your answer represented all the elements of the process above than it had each number matched with a particular step.
6. X error = .0074" above   Y error = .004" left
   Radius error = .0084"
   Geometric error burned = .0168 inches
   Tolerance error earned = .0150 (.010+.005 bonus)
   (The feature is not a hole—thus MMC is biggest version)
       MMC for this feature is   1.002
       Actual size   .997
       Bonus   .005
   This part is beyond tolerance by .0018 inches. The actual position of the center is in Quadrant 4 with respect to the perfect position.

Determining Rework Amount

Steps 4 and 5

Rework is now necessary—there is a balance due amount. Remember the amount of feature size change is either the whole balance due if the control is RFS or half the balance due if the control is MMC.

Now the new challenge after we find out how much material must be remachined is; in which direction must it be cut? We will look at that in the following section.

Example 1. The Position Control is Applied RFS (Figure 14-6)

For RFS position controls, the only option is to move the centerpoint of the incorrect hole (or feature) back within the earned tolerance along the line of error. Since the position control is RFS, the earned tolerance is simply the stated tolerance—for this example, .015 inch.

Remember the "Rule of Sides" from Chapter 13. When reworking a hole by boring sideways back along the line of error, the amount of feature size change equals the amount of error reduction. Increasing the size of a feature by, say, .010 inch, you may reduce the geometric error by .010 inch as long as the new position is inward toward perfect along the line of error. Correctly reworked, whatever change is made in the feature size can be directly taken off the geometric error. The amount of size change will be called the Feature Size Change (FSC) on the rework sheet.

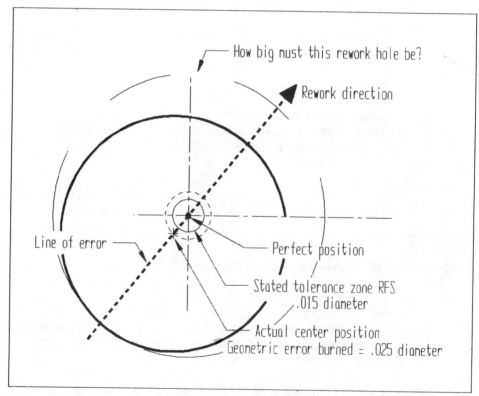

Figure 14-6 The arrow shows the line of rework direction.

CHAPTER 14 Inspecting and Reworking Two Axis Features □ 183

The FSC is required to bring the feature centerline position within the tolerance. For an RFS control, the FSC is equal to the balance due amount.

In Figure 14-6, you can see this illustrated. The solid .500-diameter circle is the original hole as machined, *out of position*. It is out of position by a .025-inch diameter geometric error.

The stated position tolerance is .015 diameter. We need to move the center of the hole back within this zone. By the Rule of Sides, how much bigger must the rework hole (dotted line) be to reduce the error circle within position tolerance circle?

Answer—The tolerance is   .015 RFS
The error is   .025
The difference   .010 Balance Due
(The Balance Due .010 = FSC because the control is RFS)

By the Rule of Sides, you must machine the rework hole to .510 inch. But it must be re-machined back along the line of error. Don't forget to see if the hole can withstand the intended re-boring amount. It is practical to go just a bit further than the minimum to ensure acceptance.

Example 2. Reworking Features Positioned MMC

The feature centerline is out of balance; however, the control is MMC. The FSC can be half the balance due. Since the position control tolerance is MMC the rework may be less difficult. You have two choices:

Option One—Reaming the Hole

First, is it possible to ream the hole up to gain enough positional bonus to cover the error? This simple solution depends upon whether the hole size has enough material left within the size range to cover the out-of-balance amount. This is on a one for one basis—.005-inch error can be corrected with .005-inch increase in hole size. The total amount necessary to balance the position error must be reamed up in the hole. For Figure 14-6, assuming an MMC control, how much must be reamed up? What size would the new hole need to be to cover the error?

Answer: .010 inch balance due. No change in position from reaming results in only gained bonus tolerance. The new hole would need to be .510 inch—in the same location but bigger to add enough bonus to balance the sheet.

Always consider this solution first but always look at the size range of the feature to determine if it can withstand that much change.

Option Two—Boring the Hole Back Along the Line of Error

The advantage here is that, for MMC controls, in boring sideways, you will gain position bonus tolerance and reduce the error at the same time.

As you learned in the previous chapter, you then need only increase the hole diameter by half the needed balancing amount.

Consider Figure 14–6. What size must the new side-bored hole be to correct an out-of-balance amount of .010 inch for the .500 diameter hole?

Answer: .505 inch. The balance will gain .005 bonus and reduce error by .005 inch.

Comparing Correction Methods RFS and MMC

Example.

Review Problem 6 of Challenge Problems 14–1 (Figure 14–5). The round boss on the hinge bracket is out of tolerance by .0028 inch. Suppose, for now, the control is RFS and it is out by .0028 inch. No more extra bonus tolerance can be gained by a size change.

Can the features positional error be corrected by moving the feature sideways along the line of error? Can the feature stand this much decrease in diameter according to the stated size tolerance? Solve this for yourself.

Answer: The diameter is .997 as machined. The smallest the feature can be is .995 inches. That is a possible FSC of .002 inch. The answer is no, the part cannot be saved. We need an FSC of .0028 while we can only reduce the feature by .002 inches.

Now suppose the control is MMC as stated on the drawing. Recall that we now gain bonus at the same rate as we change the size while reducing the error. Can the part now be saved by rework?

Now the FSC amount needed is .0028/2 Since we gain tolerance while reducing error, the feature can withstand a reduction of .0014 inch. It can be saved.

Now another question for Figure 14–5. With the MMC in effect, could we just leave the centerpoint alone and grind the diameter of the shaft to gain more bonus? (The outside counterpart of reaming, this leaves the feature in the same position but gains bonus tolerance.)

Reducing the diameter would increase the earned tolerance due to the extra bonus but not reduce the positional error. We can only reduce the diameter by .002 and the out-of-balance is .0028 inch. This method is not available either.

Neither reducing the error nor increasing the bonus alone could solve the rejection, but when the rework does both by moving the boss sideways along the line of error and changing the diameter. The part can be saved. This depends upon an MMC positional tolerance.

CHAPTER 14 Inspecting and Reworking Two Axis Features   □   185

Challenge Problem 14–2

Answers follow.

1. Compute a balance sheet and rework recommendations in Figure 14–7 for the Flange Plate. Can the rework be done by reaming or must the hole be side-bored?

   Inspection Data

Hole	X Dimension	Y Dimension	Diameter
A	.503	3.503	.312
B	1.993	2.006	1.996
C	3.491	.503	.3094

2. Make a balance sheet and rework recommendations for the Transfer Case Cover in Figure 14–8.

   Inspection Data

Hole	X Dimension	Y Dimension	Diameter
A	.9977	4.501	.497
B	3.246	2.5034	2.0025
C	6.0004	.9994	.497

3. In Figure 14–8, what if both position controls were applied RFS?

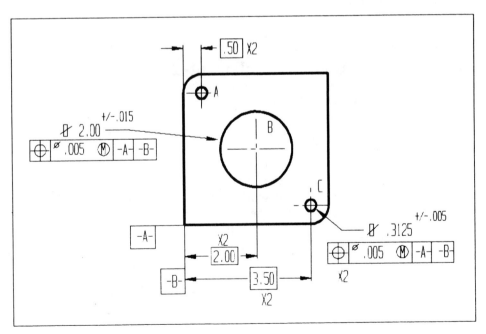

Figure 14–7 Flange plate.

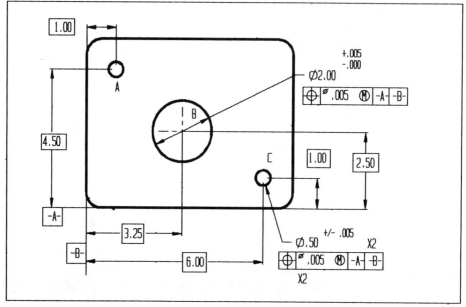

Figure 14-8 Transfer case cover.

Answers To Challenge Problem 14-2

1. Hole A Burned = .0085
   Earned = .0095
   Acceptable as is.
   Hole B Burned = .0184
   Earned = .0160
   Balance Due = .0024"

   A reamer would work. This would be an FSC of .0024 inch since there will be no change in position. We would need a 1.9984-inch reamer. However, a 2.000-inch reamer would also be acceptable and more common.

   Hole C Burned = .0190
   Earned = .0069
   Balance Due = .0121

   Reaming is not possible. We must side-bore this one feature. FSC is .0061 (take a bit more.) The new hole will be .3155 inch and moved toward perfect along the line of error.

2. Hole A   Burned = .0050
           Earned = .007
           Feature is OK.
   Hole B   Burned = .0105
           Earned = .0075
           .003 Balance Due

    Hole B cannot be reamed up .003 because the feature size will not permit it. FSC = .0015
      The hole must be bored up .0015, the FSC amount.
   Hole C   Burned = .0014
           Less than the stated tolerance—feature is OK.
           Earned = .007

3. The part is scrap—it cannot be saved.
   Hole A   Burned = .005
           Earned = .005
           Just within tolerance.
   Hole B   Burned = .0105
           Earned = .005
      This is out by .0045 and the feature cannot stand that much machining.
   Hole C   Burned = .0014
           Earned = .005
           Acceptable as is.

Determining the Rework Direction and Amount

At this point (Step 5) in the rework process, you are ready to re-machine the feature and repositioning is called for. You must know how far and in which direction the repositioned hole must be moved from the original mispositioned hole. This entails moving the centerline directly toward the perfect location along the line of error. The distance and direction to move are the problem.

Now, you need to find the real distance the feature center must be moved.

At this step in the rework process, that needed amount has already been determined—the Feature Size Change amount (FSC). It will be different depending upon whether the control is MMC or RFS.

This amount of radial change back toward the perfect position is the RC distance shown in Figure 14-9A. The RC distance is the actual physical distance the center position must be moved to restore balance. This illustration shows a radial change for an RFS position control. If the control was MMC, the RC would be less because of the bonus expansion of the earned tolerance circle.

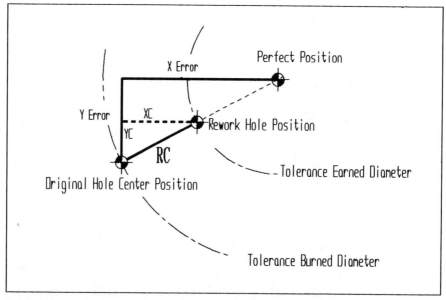

Figure 14-9A RC is the required error radial change distance.

Study Note. This discussion will be limited to reworking holes; however, the concepts are also applicable to the centerline position of other features that have round tolerance zones.

The RC Distance is *Half* the Feature Size Change Amount

Recall that, by the Rule of Sides, to change the geometric position of a feature such as a hole by .020 inch, the hole size must be increased by the same amount—.020 inch. How much does that actually move the centerpoint of the hole? The answer is .010 inch. The centerpoint of the hole actually moves one-half of the change in diameter.

The distance RC, then, is the actual shift of the centerpoint of the hole, toward perfect along the line of error. The RC distance will be half the amount of feature size change. So, for example, if the FSC is .006 inch, then the RC amount will be .003 inch.

RFS Example (Figure 14-9B is extracted from Figure 14-9A.).

The outer circle represents the actual position of the incorrectly positioned hole. The inner circle is the tolerance earned; for example, .010 tolerance earned. The

CHAPTER 14 Inspecting and Reworking Two Axis Features □ 189

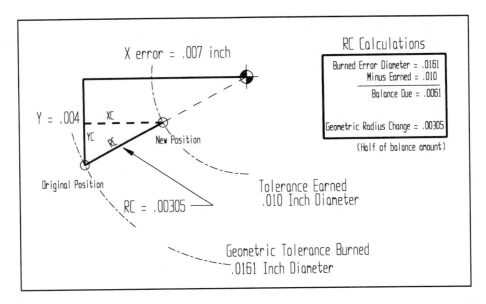

Figure 14–9B The X change and Y change distance.

actual position is calculated from the inspected X and Y distance from perfect or from a CMM readout.

After this feature was inspected it was found that it had an X error of .007 inch to the left and a Y error of .004 inch down from perfect. Through Pythagorean's Theorem we then determine that the error *diameter* is .0161 inch.

In Figure 14–9B, .0161 Error–.010 Earned = .0061-inch Balance Amount. We need to reduce the error by .0061 inch to bring the feature within tolerance balance. This is an RFS control so there is no bonus. Therefore, the FSC will equal balance amount.

From the Rule of Sides, we know that we then must take .0061 inch off one side of the hole to reduce this much error. The objective is to reposition the hole by boring the diameter larger—in the correct direction—to reduce the position tolerance burned to within the tolerance earned circle.

Determining the direction (which side of the hole) and the method of change remains to be done. The key is the RC line. RC is the actual distance change you are about to make to reposition the hole.

RC lies in the small triangle XC, YC, RC. There are two possible ways to move to the new position required for balance. You could make a small XC (X correction) and YC (Y correction) move from the original hole along the line of error. This is the *rectangular method*. Or you could rotate the part by the GA angle

(Geometric error Angle) then move inward along the line of error by the RC amount. This is the *polar method*.

Rectangular Method

This method involves finding the XC and YC small correcting distances. It is for machinery such as standard milling machines where the rework will be accomplished with a dialed X and Y distance to reposition the hole.

Polar Method

Moving the RC distance at angle GA along the line of error is the polar method. This method is where it is possible to rotate the part to align the line of error with a machine axis or to move at a selected angle for a distance such as on a CNC bore mill.

The method you select depends upon the type of machinery upon which the rework is to be performed. We will examine each method more thoroughly. The polar method is more direct and less prone to error; however, the rectangular is more common due to the types of machinery upon which rework is usually accomplished.

Rectangular Method of Correction

Here we must calculate the XC and YC correcting moves. These small XC and YC amounts are with *reference to the present hole location*—in other words, how far right or left, up or down on the part must we move the center of the hole to bring it within tolerance. This is known as the relative quadrant, with reference to the present hole position.

This problem could be solved using Trigonometry, but we will use ratio and proportion because it is simpler. We need to calculate mini-reverse versions of the original error (see Figure 14–9B). The moves are from the original hole inward toward perfect along the line of error.

These distances are found with the following *Rectangular Correction Formula*.

Notice that we use the original RE distance. This is the large hypotenuse in Figure 14–9B. Recall that the RE is the Radial Error distance from the Perfect Position to the Original Hole Position. We need the RE to compute XC and YC amounts. You solved for the RE when computing the error burned.

$$XC = \frac{X \times RC}{RE}$$

$$YC = \frac{Y \times RC}{RE}$$

where:
    RE = Original Geometric Radius Error
            The larger hypotenuse in Figure 14–9A
            (the geometric error divided in half)
  X or Y = Original X and Y errors
    RC = Change in Geometric Error Radius
            The small hypotenuse in Figure 14–9A
            Half the amount of change in the hole size.

Rectangular Example Problem (Figure 14–9B).

To reduce the geometric error by .0061 inch, according to the Rule of Sides, we must increase the hole size by .0061-inch FSC.

$$RE = .00805 \text{ (the original radius of error)}$$
$$RC = .00305 \text{ (half the FSC)}$$
$$X = .007 \text{ (original X position error)}$$
$$Y = .004 \text{ (original Y error)}$$
$$XC = \frac{.007 \times .00305}{.00805} = .0027$$
$$YC = \frac{.004 \times .00305}{.00805} = .0015$$

These are the small incremental moves from the original hole center to the new hole along the line of error. By moving the boring machine this much in each axis, the new position would be back along the line of error exactly half of the intended change in hole size. Now you must bore the hole to the new size—.0061 inch larger.

The Quadrant is Important

You must also realize in which quadrant the error lies with respect to perfect, so that the XC and YC moves will be in the correct but reverse direction to the error.

    In practice, you might choose to take just a bit more than the absolute minimum amount needed for balance. To check your work you may apply Pythagorean's Theorem to the XC and YC amounts. Taking the square root of the sum of the square of both XC and YC should equal RC.

$$RC = \sqrt{XC^2 + YC^2}$$

Challenge Problem 14-3—Finding Rectangular Corrections

Answers follow.

Rework Form

Figure 14-10 is a Rework Form that will organize your work and guide you through the process. This form starts from the time when a balance due amount has been determined. It then guides you to make decisions about the kind and amount of repositioning. There are several key numbers that must be filled in as you go:

- Original X Error
- Original Y Error
- Original Radial Error
- Balance Due
- Feature Size Change
- Radial Change

Then the rework computations:
Based on the Radial Change:

- XC and YC (if you select rectangular correction methods)
- GA and RC (if polar methods are used)

The FSC, RC, and related rework numbers will differ depending on whether the control is RFS or MMC. There are two extra forms in the Appendix of this book. Photocopy several of these and save a master for future use.

1. Calculate a complete balance and rework analysis of the .75-diameter hole in Figure 14-11. Start with the original X and Y errors. Use a rework sheet to organize your work.
   A) Compute earned positional tolerance.
   B) Compute burned tolerance.
   C) Compute the feature size change needed to restore balance. Notice that the control is MMC. Half of the balance will be gained from additional bonus when reboring.
   D) Can the feature size withstand this much change?
   E) Sketch the problem and rework on the sheet.
   F) Can the part be reamed on present location?
   G) Calculate the XC and YC correction amounts.
2. Complete a Rework Sheet for the part in Figure 14-12 as machined. Would it be practical to ream this feature? Can the hole withstand the ream FSC? Compute the XC and YC corrections.

CHAPTER 14 Inspecting and Reworking Two Axis Features   □   193

GEOMETRIC POSITIONAL REWORK

Correction Facts

What is the Balance amount needed? _____

What is the feature size change? _____

Radius change is 1/2 of feature size.   RC=_____

Feature C/L   Feature C/L
X Error = _____   Y Error = _____   Geo Radius = _____ GR

Correction will be: [ ] Polar   [ ] Rectangular

Polar Method

GA=   RC=

TAN GA=Y/X

SIN GA=Y/GR

COS GA=X/GR

Rectangular Method

XC=   YC=

XC= X x RC
    ―――――
     GR

YC= Y x RC
    ―――――
     GR

Rework Sketch

Sketch Error and Corrections in Correct Quadrant

Figure 14–10   Rework worksheet.

194 □ Unit III Application Skills in Manufacturing

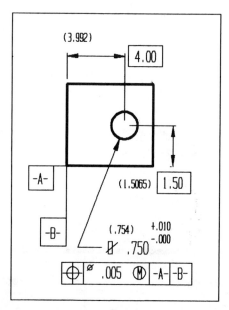

Figure 14–11 Find rework recommendations.

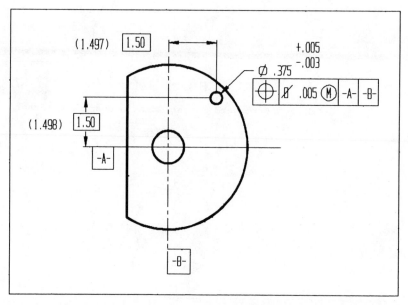

Figure 14–12 Solve the rework problem. Position details for datum -A- omitted for simplicity.

CHAPTER 14 Inspecting and Reworking Two Axis Features □ 195

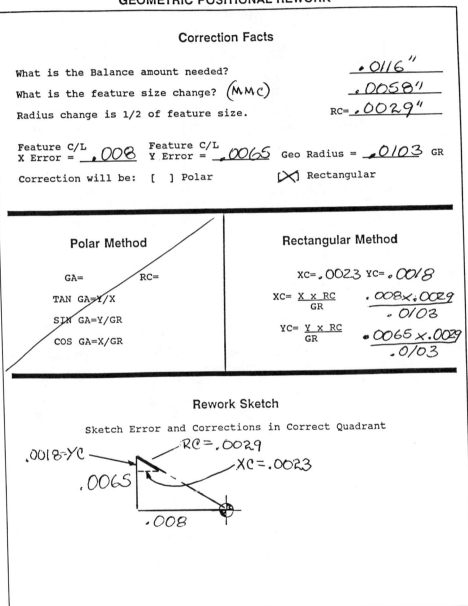

Figure 14–13 Rectangular rework sample.

Answers to Challenge Problem 14-3

1. For Figure 14-11 (also see Example Form, Figure 14-13). This is an MMC control.
   A) Earned = .009"
   B) Burned = .0206"
   C) Balance Amount = .0116"
      (MMC) FSC = .0116/2 = .0058 inch
   D) Error is in Quadrant #2 from target
   E) This part cannot be reamed to gain enough bonus. That would require .0116 total bonus gain—the feature size could not stand that much increase.
   F) The new hole will be .754+.0058 = .7598 inch
      This is just within tolerance—rework is possible.
   G) RC = .0029   X = .008   Y = .0065
      GR = .0103   XC = .0023   YC = .0018
      Rework will be in relative quadrant 4 from the error position.

2. Figure 14-12
   This is an MMC control.
   A) Earned = .005"
   B) Burned = .0128"
   C) Balance due = .0058" (FSC if reaming)
      (FSC = .0029 if boring)
   D) Error in quadrant 3 relative to target
   E) Part could be reamed but it would be risky as the reamer would be just .0002 inch under the maximum size for the hole.
      The new reamed hole would be .374 +.0058 = .3798"
      Also, the reamer size would be difficult to find.
   G) RC = .0015"   XC = .0012   YC = .001" (actual YC = .00094)
      The new bored hole would be .3769 inch.

Polar Method For Correction Amounts

With this method, we are executing one of two repositioning procedures but both depend upon knowing the radius and angle of error rather than XC and YC. The correction will be made with a small angular-radial move back along the line of error. This move will be the RC distance. Since the geometric radius of error (RE) is already known from the tolerance burned computation or from the CMM, it is a simple matter to compute the angle for the line of error.

Method 1. Aligning the line of error to a machine axis—Rotating the part by the GA (geometric angle). We then bore the new hole center inward toward perfect along the now, aligned line of error—by the RC amount.

CHAPTER 14 Inspecting and Reworking Two Axis Features □ 197

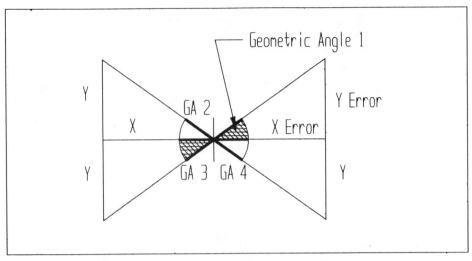

Figure 14-14 The geometric angles are numbered from the quadrant in which they lie. Each is referenced to the basic target position for the feature.

Method 2. Polar movement of the machine—Moving the boring spindle at the GA angle from the original hole toward perfect. This movement is equal to the RC amount along the line of error. This requires a programmable machine capable of angular table movement.

The key to the polar method is to calculate the geometric angle of error. A polar move requires an angle and radius distance. The radius distance is already known; it is the RC. You may compute the GA using any one of the following formulae, using original X and Y errors and the RE.

1. GA = INV-TAN Y/X
2. GA = INV-SIN Y/RE
3. GA = INV-COS X/RE

The X and Y values are always positive for the purpose of computation. Also, the notation of "INV" in front of a trigonometry function indicates "the angle whose trig ratio is:". This will be the second form of each trig button on your calculator. To access an angle corresponding to a ratio requires the second function of the trig button.

The selection of one of the above formulas is arbitrary but select the information that is most accurate in the problem. Often the RE is a computed amount based upon the X and Y errors. If so, use formula 1 since it uses

198 ☐ Unit III Application Skills in Manufacturing

more basic information. When inspecting on a CMM, it is possible to obtain different combinations of sides for the angle calculation. Use the data that is most reliable.

Per the formulas, the angle calculated is shown in each quadrant in Figure 14-14.

The angle calculated with the above formulas differs depending upon the quadrant in which the error lies, but the angle is always adjacent to the X axis (left and right) of the problem.

Challenge Problem 14-4

Answers follow.

Solve rework amounts using the polar method.

Find the RC and the GA to bring the feature within position tolerance balance.

1. Complete a polar rework sheet for Figure 14-11.
   Calculate the RC and GA (angle of rotation for rework).
   See answer—Figure 14-15—Rework Sheet
2. Complete a polar rework sheet for Figure 14-12
   Find the GA and RC for reworking the .374 hole.

Answers to Set 14-4

1. For Figure 14-11
   (Also see Figure 14-15—sample polar rework sheet)
   GA = 39.0939 degrees
   RC = .0029 inches (FSC=.0058 inch)
2. For Figure 14-12
   GA = 38.6598 degrees
   RC = .0015 inches

Note: In both problems, the Tangent ratio was the correct method to compute the GA because the X and Y original errors were the reliable data.

CHAPTER 14 Inspecting and Reworking Two Axis Features ☐ 199

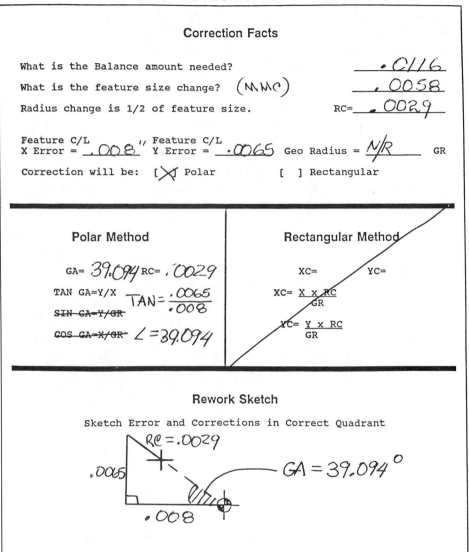

Figure 14–15 Sample polar rework sheet.

Review of Rework

Congratulations. You now have a working ability of a valuable shop skill. Rework of mispositioned geometric features follows the five-step process for any control—single or double axis. This ability can save thousands of dollars on otherwise useless parts.

First Step: GE Find geometric tolerance error burned.
Second Step: TE Find the geometric tolerance earned.
Third Step: BA Determine the balance.
If earned is greater than or equal to burned, the part is acceptable.
If earned is less than burned, consider rework.
Fourth Step: FSC Determine amount of feature size change required to correct the balance amount due.
Is the control RFS or MMC?
Determine the feature size change amount.
Fifth Step: Rework Computations Determine if the size of the feature will permit this much FSC without going beyond size tolerance.
Find the RC distance (1/2 the FSC).
The feature centerline must be moved by the RC distance. You need to know in which direction from the original location the movement must be made. Which quadrant? How much?
Compute the rectangular XC and YC movements. **OR**
Compute the polar GA and RC movements.

— Balance Amount Due. The difference in earned tolerance and the burned position.
— Is the Control MMC or RFS? If the control is RFS, then the balance amount will be entirely made up by repositioning the hole by boring. If MMC, half the balance amount will come from more bonus and half from repositioning.
— An MMC control means that the FSC can be half BA. If the machined feature size permits, MMC reaming is also possible as long as the size can withstand the entire BA.
— Can the Feature Size Withstand That Much Change? Will the change in feature size required for balance leave the feature within tolerance? If yes, then the rework is possible. If no, then the part is scrap.

CHAPTER 14 Inspecting and Reworking Two Axis Features □ 201

— Determine the RC Distance. From the feature size change amount, we divide by two to derive the radial change distance. The RC is the physical distance we need to move the feature centerline, toward perfect, along the line of error. The RC is half the FSC increase. This remains true for single or double axis controls.
— Calculate the XC-YC or GA-RC Rework Movement. You will select either polar or rectangular methods to reposition the machine over the new hole location. Use the rework sheet to stay organized. This will yield either an XC-YC correcting pair or a GA-RC pair.
— Incremental Reverse Moves. These pairs of correcting moves will produce a centerline change equal to the RC distance from the old position to the new position. This RC movement will occur in reverse direction to the original line of error—back toward perfect along the line of error.

Note: If the position control is a single axis such as a tab or slot, then the correcting moves will simply be along a single axis back toward perfect.

CHAPTER 15

Computing Geometric Tolerances for Designs

This chapter will introduce some basic methods used in dimensioning and tolerancing new designs. The understanding and oversight gained here will enable you to better understand geometric prints and where the tolerances originate. As a shop person, you will better understand the how and why of the tolerances and controls you find on prints.

Geometric Priorities Linked to Function

Prior to tolerancing a drawing, at the very start of any design, the designer must compute the allowable deviation in any assembly. The fit and function of the assembly or individual component will determine this in most cases. This functional analysis must precede the dimensioning and tolerancing. To receive the full benefit of the system, the geometric advantage is built into the design from inception where the tolerable allowances are derived based upon the function of the part or assembly.

The entire process begins with a geometric analysis of the product function. How tight and how loose must the selected feature control be? Which features or function is the most critical? The least? Engineering decisions must be made about the range of acceptability for each feature of the design.

Take for example, how much backlash must be in the system or how tight the press fit amount; the maximum gap between assembled components; mismatch, alignment, or form deviations such as flatness. All of these design requirements must be researched before the geometric dimensioning and tolerancing can be applied.

Once this range of deviation is determined, the drawing may be dimensioned and toleranced, with the objective being to use the entire natural tolerance available. Deriving and using the greatest amount of tolerance possible holds costs down while making the largest range of correct parts that will function per design.

Although the process is complex at times, using geometric analysis helps to see the true tolerance available and how to portion it out between various com-

CHAPTER 15 Computing Geometric Tolerances for Designs □ 203

ponents of a design. A geometric analysis has another big advantage: dividing out critical features to the design from the less critical. This assists total quality control of the product by deriving the best control points in the process. A geometric analysis of the design then allows the designer to assign close tolerances where needed and to loosen the tolerances when possible. Statistical process control (SPC) is much more effective if the process is serving the geometric function of the design needs.

Recall that in Chapter 1 we discussed the difference in a geometric and nongeometric design where the flatness of the top of a block was being controlled by the height tolerance. Recall that on block number two (the geometric toleranced one), the geometric block was going to be much easier to make and far less prone to scrap because the flatness was controlled geometrically. The geometric separation of the controls for height and flatness then provides far more control of the manufacturing process. These are the control points needed to reduce deviation through SPC.

Another example is making a feeder plate to send sheets of paper into the inlet of a printing press (Figure 15-1). Let's examine what is truly critical about this assembly. For discussion, top priority is given to the fitup line where the feeder plate matches the next plate in the press. The paper must slide past this joint in the two plates. The joint itself must not protrude and be straight to avoid catching the paper. Let's assume .0004-inch maximum mismatch between the two components. This natural tolerance for mismatch could be arrived at several ways: previous experience, an engineering handbook, or experimentation. Any more mismatch would catch or scratch the paper sheets. Due to the close tolerancing, this will be an expensive pair of components but to hold the expense

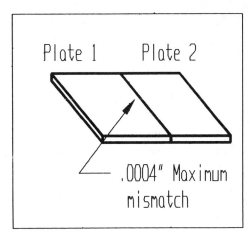

Figure 15-1 Controlling the maximum mismatch between two plates.

down we need to get the entire natural .0004 tolerance into the design. Also, we need to be conscious of the way in which we dimension and tolerance the fit.

Would flatness be the best control for the design? Perhaps we could use straightness parallel to the joint in the two plates? Straightness is actually the major problem at the joint. Because straightness is a single element control it is easier to machine and inspect than flatness, thus a less costly control.

Tolerancing the design, we cannot put .0004 straightness tolerance on each component in the assembly. Why not? Because the resultant maximum mismatch between the two could be .0008 inch. We must "spend" the tolerance requirement between the two components; that is to say, decide how to portion the available tolerance between the components.

The spending requires some further investigation. Which component will be the more difficult to machine? If there is no difference, the .0004-inch total may be divided equally between both parts. Each part might have a straightness tolerance of .0002 inch. However, if one part is more difficult to machine, perhaps it is a thin casting prone to movement from internal stress. It could receive .0003-inch straightness control and the other component .0001 inch. The resultant mismatch would then never exceed .0004 inch for the assembly.

Any combination yielding .0004 total tolerance between both components is acceptable per function. From the cost standpoint, various combinations of the tolerances are going to have different manufacturing costs.

The process then begins with a geometric/engineering look at the function of the part or assembly. How much deviation is allowable? Only then are you able to use geometric methods to dimension and tolerance the drawing.

The investigation skills and methods whereby we arrive at the natural fit and function tolerances, the amount of clearance in an assembly, are not the purpose of this book. In this chapter, you will gain a background knowledge of the reasons for the tolerances you need to achieve in the shop. We will concentrate on the geometric methods for deriving the actual tolerances, after the engineering decisions have been made.

Virtual Condition

The first concept used to examine an assembly is "virtual condition." This is the state of an assembly or individual component whereby the worst possible state is tested for acceptability of design and tolerance. Virtual condition is the net effect of all tolerable deviations to the final resultant geometry.

A simple example are the pair of bridge beams held together with two bolts in Figure 15-2. We will soon see that the tolerancing for position of these two components must be calculated using a process called "floating fastener."

For now, consider the two extremes of assembly. What are the tightest and

CHAPTER 15 Computing Geometric Tolerances for Designs   □   205

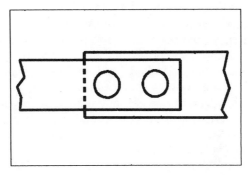

Figure 15-2 Using the engineering facts, calculate the maximum and minimum hole size and the position tolerance.

loosest assembly conditions that are acceptable? Keep in mind that the tightest state MMC must still have some clearance for assembly and also accommodate heat expansion of components in actual use of the bridge.

Due to the calculated shearing action load on the bolts, we will choose a 1-inch bolt. The actual factory-specified size of the bolt is .990 inch. The manufacturing tolerance for the bolt diameter is plus or minus .005 inch. The bolt will vary between .985 and .995 inch. That starts the analysis process. We must analyze the assembly with the largest (MMC) and the smallest (LMC) bolts and include the position error and clearance factor too.

The example maximum acceptable clearance is .050 inch between bolt and hole. Let's assume that the tightest fit we can withstand is .010 clearance between the bolts and hole.

Bridge Beam Engineering Facts

Bolts
    MMC .995"
    LMC .985"
Clearance
    MMC .010"
    LMC .050"

Question? What could the smallest and largest holes be in the beams? Calculate the maximum and minimum hole size to accommodate the engineering facts above.

The smallest hole possible in the assembly would be found by looking at the MMC factors. This would yield the tightest fit. A .995-inch bolt and .010 clearance

= 1.005-MMC hole. The largest hole possible is found at LMC for both the bolt and hole—.985 + .050 clearance = 1.035 inch.

The holes may range between 1.005 and 1.035 inch. Next we must derive the positional tolerance for the holes. In this example, the positional tolerance will be calculated using a method called Floating Fastener due to the fact that the bolts may self-align to the overall resultant holes.

The overall never-violated hole available in the *assembly* is the result of the size, position, orientation, and form and related tolerances in each component. All these factors for the drilled holes must be chosen so that the two beams will always assemble at both extremes of manufacturing. The beams must assemble at the tightest MMC Virtual Condition and loosest LMC condition but not exceed the looseness limit.

At assembly, the dimensioning and tolerancing must ensure a cylindrical zone that remains open through both parts, through which the bolts always fit. This reserved zone must not be smaller than the bolt MMC size. This unviolatable cylindrical space shown in Figure 15-3 must be no smaller at any point than the MMC fastener. In other words, for all possible conditions of the manufactured product, this minimum reserved cylinder must remain open through both beams for each mated hole.

You can see why perfect form at MMC becomes important in this example. Perpendicularity could also play a part if the beams were thick because the reserved cylinder produced from the combined holes must be straight enough to accept the bolts.

The purpose of the controls selected, material modifiers, and tolerancing is to

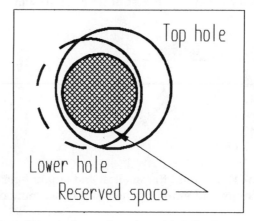

Figure 15–3 Virtual condition is the combined result of all tolerances. When both features are their extreme limits for position, form, and tolerance, this reserved space must exist to allow assembly. The design tolerances must ensure this.

ensure assembly at virtual condition and MMC. With the holes at their smallest MMC state, and the bolts at their largest MMC condition, plus position error at the limit of tolerance, the minimum reserved cylindrical assembly space must exist.

Calculating Position Tolerance Using the Floating Fastener Tolerancing Method

The beam illustration above is an example of one of two tolerancing methods for fastener assemblies. The floating method is so named because the fastener has the freedom to align to the reserved cylinder space in whatever attitude it is found. Floating assemblies have a calculated amount of freedom to move.

We now know all the engineering facts necessary to derive the position tolerance. The position tolerancing of the holes is based upon two decisions:

The Bridge Beam Engineering Facts

1. The MMC and LMC size of the fastener—bolt in this case.
   .995" MMC and .985" LMC
2. The largest and smallest hole possible.
   1.005" MMC (.995" + .010" min clearance)
   1.035" LMC (.985" + .050" max clearance)

All engineering decisions having been made, the following formulae apply in dimensioning and tolerancing the holes in the beams.

$$PT = HMMC-FMMC$$
Positional Tolerance = Hole Size MMC—Fastener MMC Size

The maximum material condition geometric position tolerance for the holes in the bridge beams then is:

$$1.005-.995 = .010 \text{ inch.}$$

At the MMC hole size, each hole centerpoint must lie within a round tolerance zone of .010-inch diameter. If this condition is met, the beams will assemble per design function. If the holes deviate away from MMC, the positional tolerance may grow.

Each Component Receives the Full Positional Tolerance

Notice also that both bridge beams would have the full MMC positional tolerance for the holes. Because the fastening method is floating, there is no need to divide the tolerance between both plates. Each may be toleranced at .010 inch. This is only true for floating fasteners because of the self-alignment of the fastener.

Maximum Material Bonuses Apply

Notice also that the bonus advantage concept applies with floating fastener assemblies. As the holes deviate away from the MMC size, the position tolerance may grow in direct proportion.

Question?
What is the largest position tolerance possible in this design?

Answer:
MMC size = 1.005
LMC size = 1.035
Difference .030 inch Possible Bonus
At LMC, the holes could have a positional tolerance of .040 inch.

Question?
What is the largest clearance space between a bolt and hole that might occur on one side of a hole?

Answer:
LMC hole = 1.035"
LMC bolt = .985
Difference .050 inch

This is due to our original design decision that the maximum clearance was to be .050 inch. If an LMC bolt is held to one side of the LMC hole, the largest space can be .050 inch.

Challenge Problem 15-1

Floating Fastener Assembly Position Tolerances
Answers follow.

1. Design the three holes to fasten an access port hinge to a navigation computer door (Figure 15-4). We are to use 1/4 inch rivets for which the production tolerance is .240 to .245 diameter. The fasteners must fit with a minimum clearance of .005 inch and a maximum of .015 inch.
   Complete the figure by:
   A) Finding the floating fastener MMC positional tolerance for this assembly.
   B) Expressing the hole size and tolerance.
   C) Finding the maximum positional tolerance possible for a single hole at LMC size.

CHAPTER 15 Computing Geometric Tolerances for Designs □ 209

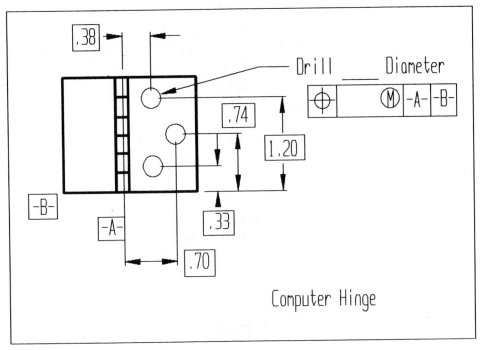

Figure 15–4 Complete the size and position tolerance.

2. Calculate the positional tolerance for drilling holes in a fuel cell baffle. It is to be riveted into the assembly with .100-inch diameter rivets that have a tolerance of +/− .002 inch.
Minimum clearance is .003 and the maximum is .010 inch.
Notice that each component received the full calculated tolerance.

Answers to Challenge Problem 15–1

1. The engineering facts are:
   A) Fastener = .240 LMC
              = .245 MMC diameter
   B) Hole MMC = .250 inch (MMC bolt plus .005 min. clearance)
         LMC = .255 inch (LMC bolt plus .015 max. clearance)
   The floating positional tolerance is found with the above formula.
   PT = Hole MMC−Fastener MMC

PT = .250 – .245 = .005 position tolerance MMC.
The hole size and tolerance can be expressed several ways:

I) Unilateral: .250" + .005
                       – .000

II) Bilateral .2525" + .0025
                       – .0025

III) Limits .250"
             .255"

The maximum positional tolerance would be at LMC hole size. .005 + .005 bonus = .010 inch

2. Facts:
   Fastener MMC = .102"
            LMC = .098"
   Holes MM = .105 (.102 + .003)
          LMC = .108 (.998 + .010)
   Positional Tolerance = HMMC – FMMC
                        = .105 – .102 = .003 inch at MMC.

Calculating Positional Tolerances Using the Fixed Fastener Principle

A fixed fastener assembly is designed such that one component of the assembly has fasteners, dowel pins, or threaded holes in a rigid position. In Figure 15-5, you see two examples of fixed fastener assemblies. The first has two protruding pins that must fit into a mating part. The second example is of a threaded part over which a mating plate must fit. When the bolts are screwed in, they form a rigid location for the fasteners.

Positional Tolerance Divided Between Mating Components

In any fixed fastener assembly, the total assembly positional tolerance is calculated with the same position formula as for floating fasteners; however, this amount must be divided between the two components. It need not be divided equally and may be portioned between the mating components based upon the machining difficulty of the mating features. In other words, the more difficult feature of the two mating parts may have more of the position tolerance while the easily machined feature receives the lesser tolerance amount. But the total must add up to the amount derived from the tolerance formula:

Position Tolerance = Hole MMC – Fastener MMC

CHAPTER 15 Computing Geometric Tolerances for Designs □ 211

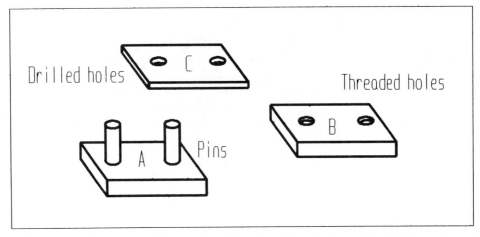

Figure 15–5 Plate C assembles to either Plate A or Plate B using a fixed fastener concept.

The available calculated position tolerance is divided between each component—the fastener plate and the mating plate. The component with the rigid fasteners receives a portion of the tolerance and any mating components then receive the remaining position tolerance. If several components must mate to a rigid fastener plate, then each must have the same tolerance as the first. In so tolerancing the assembly parts, all mating parts will reserve the cylindrical space necessary for fit.

Example.

In Figure 15–6 we are to mount a plate over a mold base. We will bore and press fit a pair of pins in the base and bore a pair of corresponding holes in the top plate. In all conditions, the top plate is to slip over the base pins. What is the positional tolerance for the bored holes in the base and in the top plate?

First, the engineering facts derived from a tooling handbook.

Pins
   MMC .998"
   LMC .997"
Clearance—from a similar previous product.
   MMC .003"
   LMC .006"
What size range must the top holes fall within?
   MMC Hole = MMC pin + MMC clearance = .998 + .003 = 1.001 inch
   LMC Hole = LMC pin + LMC clearance = .997 + .006 = 1.003 inch

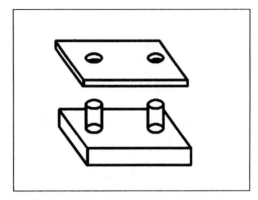

Figure 15–6 Calculate the positional tolerance for the bored holes in each plate.

The position tolerance formula then yields MMC Hole−MMC Pin
1.001−.998 = .003 inch total position tolerance

Remember, as a fixed fastener assembly, the positional tolerance of .003 inch must be divided out between the two components. Since each component in this case is a bored hole, we might divide it equally. The holes in the base and the holes in the top plate each have .0015-inch position.

Meeting the .0015 positional tolerance and the engineering facts as well, in all states of virtual condition, this pair of components will assemble. We could have divided the available .003 positional tolerance in any ratio as long as the two amounts add up to .003 inch. In actual practice, perpendicularity of the bored holes in the base will also be a factor. A projected tolerance zone would be used to control the axis of the holes to the length extent of the intended press fit pins.

What if There Is More than One Plate to Assemble over the Pins?

In the example above, what position tolerance does each of three plates receive if they are to then stack assemble over the pins?

You should have answered that each plate receives the same tolerance as the first. As long as each plate maintains the same tolerance as the first, they will all assemble over the pins. The base gets a portion of the position tolerance and all mating parts get the same remaining portion.

Envision all the possible positions of the largest MMC pins in this design. What shape, or space, do they then take up. In other words, if we were to draw a graph of all the possible positions of the press fit pins what shape would we

CHAPTER 15 Computing Geometric Tolerances for Designs □ 213

draw? They will form the reserved cylinder (emulating MMC pins, plus location allowance) over which the plates must slide. This reserved cylinder will be imaginary pins .0015 inch larger than the MMC pin due to the given position tolerance. Holes in mating plates must not violate this round space of .9985-inch diameter.

Correctly dimensioned and toleranced, the top plate will always produce this reserved cylinder for assembly. This must account for base plate position tolerance, MMC pins, and clearance. Then any top plate hole size deviation away from MMC will yield more clearance thus bonus position tolerance could apply.

Bonus position tolerance could not apply to the base plate. Why? Because the size upon which the assembly position bonus would depend is the pin size not the bored holes. A size change in the bored hole simply changes the press fit status.

Since the pin size is only known as a size range, the engineer tolerancing the base must assume an MMC pin. In other words, if the bored hole were to deviate away from MMC, it would only change the press fit, not add location bonus. All the above facts will ensure the reserved cylinder through the entire stack of plates as they are assembled to the pin base.

Functional Gaging

We have designed a functional gage in the above discussion. The reserved pin cylinder of .9995 would test the upper plates. A functional gage tests all possible conditions of the mating part. To test the upper plates would require a base plate with pin diameters of .9995 inch on the perfect location. This gage would test for plate MMC assembly. All plates must slip over this simulation of all the possible conditions of the base pins.

A similar functional gage could be built to test the base plate that simulates the MMC condition of all the possible upper plate holes. It would have the .9995 reserved hole as the test. Any base plate pin combination must then slide into this hole gage.

Challenge Problem 15-2

Fixed Fastener Assembly Position Tolerances
Answers follow.

1. We are to design a cylinder head that must assemble to an engine block. The engineering facts are:
   A) Bolt sizes .620 to .622
   B) Clearances .025 to .010

214 □ Unit III Application Skills in Manufacturing

What is the positional tolerance available to be portioned out between the two components?

If the tapped hole position in the block tolerance was stated at .004 inch, what MMC position tolerance is available for the cylinder head?

2. Complete the feature control frame (Figure 15–7) for the printer arm. Calculate the fixed fastener position tolerance.

It is to assemble to a printer base (not shown) which has three fixed pins protruding. The position tolerance used for the location of those base pins was .001 inch.

Engineering facts:

Pin
$\overline{\text{MMC}}$ = .1875
LMC = .1865

Clearance
$\overline{\text{MMC}}$ = .0010
LMC = .0040

3. In Figure 15–7, after machining,

What position tolerance would apply to a hole of .189-inch diameter?
What is the position tolerance for a hole of .192 diameter?

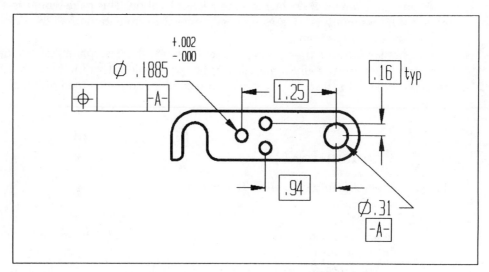

Figure 15–7 Compute the hole position tolerance. Use the fixed fastener method.

Answers to Challenge Problem 15–2

1. The position tolerance available in the head is .006 inch MMC.
   Engineering Facts
   Bolt MMC = .622"
   Hole MMC = .632"
   Available Tolerance = .632 − .622 = .010 inch.
       Since the block tolerance consumed was .004 inch then the head position tolerance may have the remaining .006 inch.
2. The position tolerance available at MMC hole size is .000" (zero)
   Engineering Facts
   Pin MMC = .1875
   Hole MMC = .1885
       Total assembly available position tolerance = .001 at MMC. Since .001-inch position tolerance was consumed by the pins at MMC, the holes have a zero position tolerance at MMC. They still may have a position tolerance but must be at some size other than MMC size to earn it. For a review of Zero Position Tolerance please turn to Chapter 11, Concept #29.
3. The hole at .189 inch would have .0005-inch position tolerance. The hole at .192 would be scrap—beyond the size limit of .1905-inch LMC diameter.

    This chapter has been introductory for craft people only. Its purpose was to provide you with some understanding of where the design tolerances you must achieve, in the shop, originate. There is much to be gained from ANSI Y14.5-M and from design experience. There are many formulas and recommendations for calculating fits and tolerances in ANSI Appendix B—"Formulas for Positional Tolerancing."

CHAPTER 16

Complex Tolerances

This is the final applications chapter where we will explore two new concepts.

1. Double bonus tolerances resulting from datums of size.
2. Complex rework.

Both are advanced skills required in real manufacturing.

Bonuses from Datums of Size

A datum is established by part features. Often the feature chosen as a basis for a datum might have some size variation such as in Figure 16-1, where the datum is the hole. On this design, the most important aspect of the slot is the horizontal relationship to the hole. The hole then becomes the datum for the slot.

This datum then is a *Datum of Size*. It has a range of acceptable sizes. As this datum varies from MMC size, it is functionally possible to gain bonus position tolerance for features that relate to it. This bonus position tolerance can apply to a single feature or to a group of features. In this example, it applies to a single feature.

Figure 16-2 shows a pair of mating parts. For now we will limit our discussion only to the plate with the holes. The part with the pins is a perfect test part and not in discussion. Functionally, in this assembly, the relationship between the holes is most important. The outside edges of the plate as compared to the holes are unimportant. In Figure 16-3, dimensions for the outside edges have been omitted for simplicity.

There are two holes in the plate. The position of the second hole is located from the first. The first hole becomes a datum for the second. You can see that if the second hole's size deviates from MMC that the position tolerance can grow and a bonus is made possible. Now, if we enlarge the datum hole, would not a pin gage fit more loosely too? The answer is yes. The position tolerance of the second hole is linked to both the controlled feature hole size and the datum hole size. Functionally, referencing a feature to a datum of size can yield a double bonus tolerance.

These size-datum bonus tolerances divide into two categories: those that pertain to a single feature and those that pertain to an interrelated group of holes or other features. We will examine each.

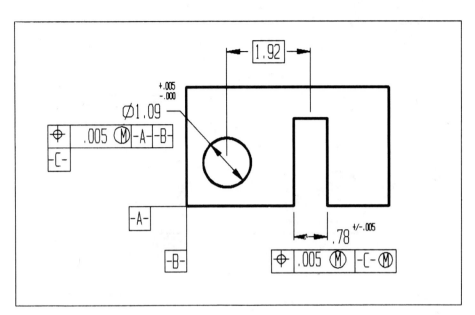

Figure 16–1 Datum -C- is a datum of size.

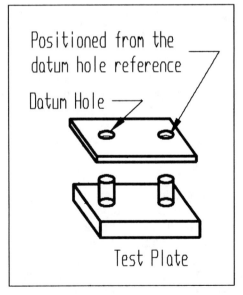

Figure 16–2 The second hole is positioned from the first, which is a datum for the position tolerance.

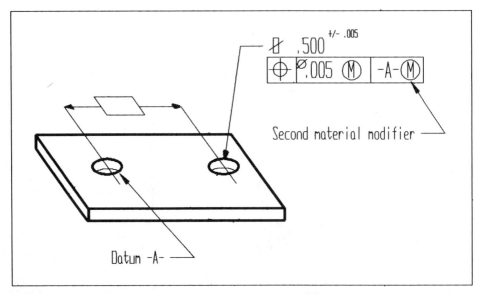

Figure 16-3 Datum -A- is a hole and must have a size tolerance. The design permits a second bonus for the position tolerance based upon the actual size of the datum hole.

The size–datum bonus possibility is signified by the placement of a material condition modifier placed adjacent to the datum symbol in the feature control frame (see Figure 16-3). The MMC symbol in the box by the A symbol signifies that it is a datum of size and as it varies from the MMC size, bonus tolerance may be earned. This bonus tolerance applies directly to the position of the feature being controlled.

Calculating Double Position Tolerance Pertaining to a Single Feature

The process for determining the earned position tolerance remains the same as you have been taught. The bonuses are only possible after the datum and controlled features are machined and measured.

There are two calculations to be made: first, the bonus earned from the datum feature size (this is a new bonus); and second, the bonus available from the controlled feature size (this we have already learned to calculate). This is not a complex calculation. It is simply finding the actual size of each feature then finding the difference from their corresponding MMC size. You get to calculate two bonuses instead of one!

Example of Calculating a Double Bonus (Figure 16–4).

The datum feature hole is produced at 1.004-inch diameter and the controlled hole at 1.003 inch. What is the position tolerance earned for the controlled hole?
Answer: .005 + .009 + .008 = .022-inch position tolerance

Calculations

Datum Feature Bonus			Controlled Feature Bonus		
MMC Hole	=	.995	MMC Hole	=	.995
Actual	=	1.004	Actual	=	1.003
Bonus	=	.009 inches	Bonus	=	.008 inches

Engineering Note: In effect, when we take advantage of this double bonus possibility, we are allowing an extra degree of movement of the features connected to the datum of size. In other words, the related features "float" within the design to the extent of their position tolerance with respect to the datum. The controlled feature is connected to the datum of size. As the datum moves and deviates from MMC, it "pulls and effects" the positional of the related feature. As the datum changes size away from MMC, the position tolerance of the controlled feature may grow.

This then implies that the relationship of the controlled feature to the datum of size is more important than the relationship of these features to other datums on the part.

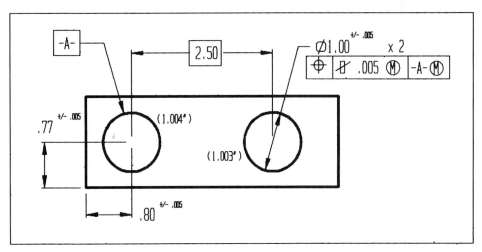

Figure 16–4 Calculate the earned position tolerance for the controlled feature.

220 ☐ Unit III Application Skills in Manufacturing

Double Bonus Applying to A Group of Features

Functionally, there can be limitations in applying the double bonus concept when a related group of features is positioned from a datum of size. Often in this situation, the overall position of the group can gain bonus tolerance as the reference datum of size deviates from MMC. But, due to the nature of the group, only the normal single bonus condition can exist inside the group.

Take, for example, an instrument panel, where we might see a compound control for position (Figure 16–5). The first control relates to the position of the whole group while the second control relates to the intergroup tolerance. In situations of this type, the bonus may only apply to the group position, not within the group. Consider the functional fit of an instrument being placed into this cutout.

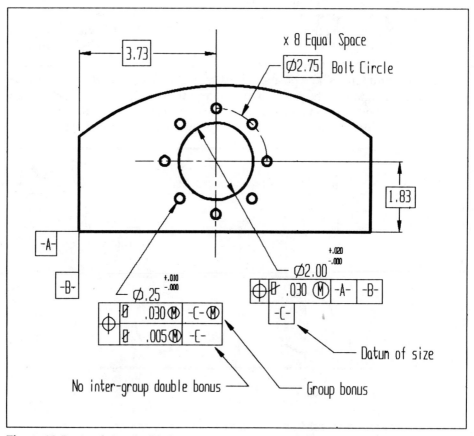

Figure 16–5 Applying double bonus to a group of features.

The central cutout is called a datum of size for the group. As this datum grows away from MMC size, the related mounting holes will have more freedom to float within the cutout; however, once a single mounting screw is threaded through a mounting hole into the instrument, we now have a fixed fastener situation and the relationship of the remaining screw locations is fixed not by the central cutout but rather by the mounting pattern of the instrument itself.

The relationship between the threaded instrument holes and the drilled mounting cutout holes is a fixed fastener one, and thus remains constant no matter what the central datum of size does. When this type of assembly occurs, there will be no second material condition modifier in the intergroup tolerance as in Figure 16-5. In other words, as the central datum grows larger, the whole group of mounting holes may float more bonus position tolerance for the group. But this is not possible within the group between individual holes.

Calculating the Size–Datum Double Bonus

In either case—1.) computing a single, or 2.) computing a group bonus—the calculation is the same.

1. Find the actual datum size.
2. Determine the datum MMC size.
3. Compute the difference.
4. Add the difference to the positional tolerance (for the group or for the single feature).

The difference lies in how this new double size–datum bonus may be used.

Applying the Bonus to a Single Controlled Feature

If computing the double bonus positional tolerance for a single feature, as in Figure 16-2, then the bonus adds directly to the position tolerance already earned for the feature. In effect, you earn more position tolerance than that which the feature alone could earn.

Applying the Bonus to a Group of Features

If computing the double bonus for a group of features, it applies only to the location of the control for the group (the top feature control frame). This means that upon inspection, the pattern position tolerance may grow as long as the datum of size is so marked with the MMC symbol in the control box (see Figure 16-5).

222 ☐ Unit III Application Skills in Manufacturing

However, the interpattern tolerance may only grow from individual feature actual size deviation. The interpattern tolerance may only have a single bonus earned from the deviation of the individual holes themself. This is as you have learned thus far. If, for example, one of the .25 diameter mounting holes deviates away from MMC, then that individual position tolerance for that hole can benefit from the single bonus tolerance.

Challenge Problem 16–1

Answers follow.

1. Calculate the position tolerance for the holes as shown in Figure 16–6. Actual sizes are in parenthesis.
2. What is the position tolerance for the slot in Figure 16–7?
3. Calculate interpattern earned position tolerance for the .253 inch hole in the upper right of the pattern in Figure 16–8.

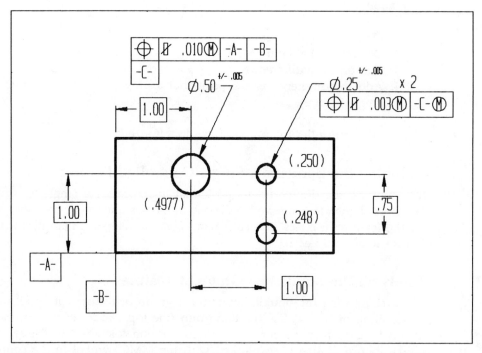

Figure 16–6 Calculate the earned position tolerance for the two .25-diameter holes.

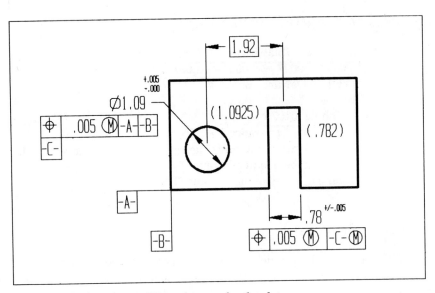

Figure 16–7 Calculate the position tolerance for the slot.

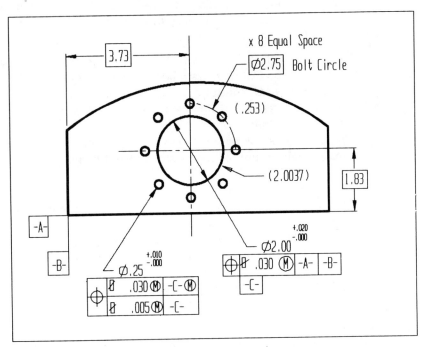

Figure 16–8 Calculate (1) group tolerance and (2) individual tolerance for the .253 hole.

Answers to Challenge Problem 16-1

1. .248 hole—Total Earned Tolerance = .0087 inch
   $\underline{.003\ +\ .0027\ +\ .003\ =\ .0087}$
   Print   Datum   Size
   Bonus   Bonus
   .250 hole .0107
2. Total earned tolerance is .0145 inch.
3. This hole may only have the interpattern tolerance plus its bonus. There may only be a single bonus.
   Total earned tolerance equals .008 inch.

Complex Rework

Often in rework situations, the relationship of one feature to another can cause complexities. Reworking an erroneous feature can cause a second to go out of positional tolerance when it was not originally.

Examine other features on the part when related to a datum of size to see what the intended rework will do to them—especially those features just within their own position tolerance. There are many ways to bring a feature into tolerance. Here is a list of the possibilities.

1. Changing the feature size without moving the center position. Machining both sides of a feature to gain bonus tolerance (gains bonus tolerance but must be an MMC control).
2. Moving the center position by remachining (reduces geometric error on RFS controls, gains bonus on MMC).
3. Equal amounts of error reduction and bonus gain (must be an MMC control).
4. Differential amounts of error reduction and bonus gain (must be an MMC control).
5. Changing the position of a datum feature.
6. Changing the size of a datum feature, thus gaining bonus tolerance for any related features (must be a datum of size and the feature in question must be related to it by an MMC control).
7. Differential amounts of datum feature centerline change versus feature size change.
8. Machining datum features that are not datums of size. For example, machining a small amount off the edge of a part to bring a drilled hole within position tolerance.

In previous rework chapters, we have considered only gaining the maximum amount of centerline movement during rework. For example, we calculated how much diameter change was required to move a centerline a certain distance thus reducing error while gaining bonus tolerance.

Differential Centerline Movement

Sometimes this centerline movement must be controlled in varying amounts with respect to the feature size change. In other words, we cannot use the full amount of centerline movement because by changing the feature position in an effort to save it, it throws some other related feature out of positional tolerance.

Occasionally, centerline movement needs to be controlled within a limit because of the effect it might have on another feature. In this case, the feature size is changed but the centerline is not moved as much as possible.

In Figure 16–9, we see an example of differential change. The solid lines represent holes that are in the wrong position. The dashed lines represent the larger size to which we can bore the hole. On the right, the change is taken as equal parts. The balance due amount is cancelled by equal parts of error reduction and gained bonus. The new hole is bored sideways to the tangency of the old. This yields the maximum amount of correction.

The hole on the left shows a differential rework. While we moved the centerline position, thus reducing some error, we did not move it as much as we gained in bonus size tolerance. This type of rework is often necessary when you are correcting a mispositioned datum of size. Suppose the hole on the left was a datum for the hole on the right? Moving it the maximum allowable amount to get it within positional tolerance might then destroy the position of the second related hole.

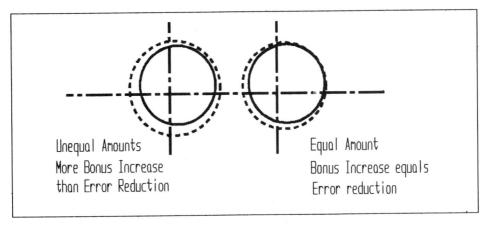

Figure 16–9 Differential rework amounts.

Costs in Complex Rework

Rework of this type is often complex and costly. Before attempting to proceed, make sure the rework is worth the effort. If there is much re-machining needed, perhaps it is more cost effective to scrap the part. As a cost saver, if you can see two methods of reworking a part, perhaps one is simpler and less costly.

The problem presented next is a good example of the complexities that can arise in rework problems. I estimate you will need an hour or more to solve it. In real shop practice, this problem might take even longer to solve. (You won't have hints and the answer available.) Therefore, depending on available material, job schedule requirements, and your time frame, you might choose to follow through on rework or simply set the part aside as not cost effective to rework.

Due to the number of possibilities for rework combinations, there are few rules or procedures to lay out for you to follow. The application skills are gained from experience. Each situation presents a new combination of rework challenges.

Challenge Problem 16–2 Complex Rework

This is the final computation problem in the book. I sincerely hope that you seldom face as difficult a problem in actual practice! Work through this puzzle one step at a time.

Instructions
A. Perform a complete inspection and rework recommendation for Figure 16–10, Transfer Case Cover.
B. There will be more than one way to rework the part to bring it within acceptability. Use the list of correction possibilities listed in this chapter.
C. You have no restrictions as to what re-machining you may do. You can machine any detail or feature of the part.
D. You are concerned only with the horizontal position of the .625-inch slot.

Hints
A. When computing the positional balance for the .625-inch slot and the .75-inch diameter hole, don't forget to add on the extra bonus gained from the datum -C- 3.000-inch hole.
B. Reworking the central datum -C- hole requires a new evaluation of the other two features. Depending on the type of rework you choose, changing datum -C- *can* effect the position of related features. Also, reworking the datum -C- hole size will add more bonus to each related feature.
C. Progressive hints will be found on the next page along with one answer for the problem.

CHAPTER 16 Complex Tolerances □ 227

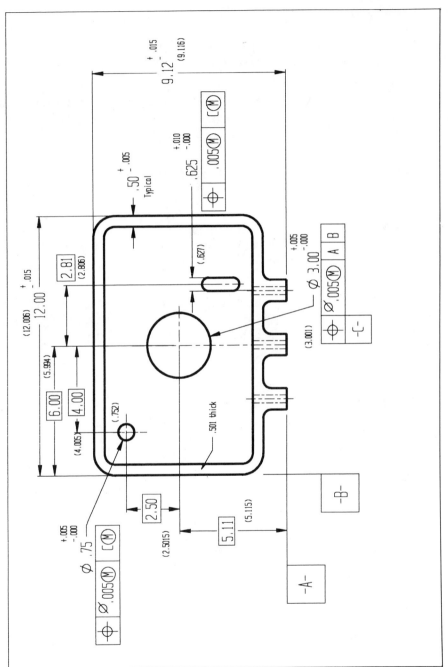

Figure 16-10 Transfer case cover.

Answers and Hints to Challenge Problem 16-2

Cover the hints and read a single one at a time.
The answer is at the end of the hints.

Hint Number 1
First, investigate what happens if you shave material off of the bottom datum -A- surface.

Hint Number 2
Shave .005 inch from datum -A-. This will not make the 9.12-inch height undersize. This will reduce the Y error to zero. Only the X error will remain.

Hint Number 3
After shaving .005 inch, the balance due for the position of datum hole -C- will be .006 inch (.012 burned and .006 earned).

That requires an FSC of .003 inch which is possible. An FSC of .003 inch produces an RC of .0015 inch.

Hint Number 4
Now re-evaluate the position of each related feature by adding in the RC of .0015 inch. Note that this .0015-inch RC is a horizontal change only. Moving the center of the 3.00-inch hole by .0015 affects the position of each related feature.

Answer to Problem 16-2
Shave .005 inch from surface A and bore up the 3.001-inch hole to 3.004 back along the now horizontal line of error. The slot will require no rework and the .750-inch hole will need either reaming of .0023 more or boring of .0012 inch back along the line of error—see facts below.

.625 Slot

Originally, the slot was .004 inch too close to the datum hole. The new datum hole will position the slot even closer. The slot's new X error will be .0055 inch when the datum rework of .0015-inch RC is added to the original shortfall of .004 inch.

Good News! After reworking the central datum, the .625 slot will just be in positional tolerance. No rework required.

Earned Calculations
MMC Size .625
ACT Size .627
Bonus .002
.002 + .004 + .005 = .011 inch earned

| Bonus Datum | Print Tol. | Bonus Tol. |

Burned Calculations

RE = .0055
GE = .011 Inch

.750 Diameter Hole

After reworking the central datum, the .750 hole will require rework and a new balance calculation. It can be reamed up .0023 inch in the same position, or bored up .0012 back along the line of error. Reaming would produce a hole of .7543 inch while boring would make the hole .7532 inch. Both solutions are acceptable.

Earned Calculations

MMC Size .750
ACT Size .752
Bonus .002
.002 + .004 + .005 = .011 inch Earned
Burned
X error = .0065
Y error = .0015
RE = .0067
GE = .0133 Inch Burned
Balance Due = .0023 inch

UNIT IV

Geometric Theory

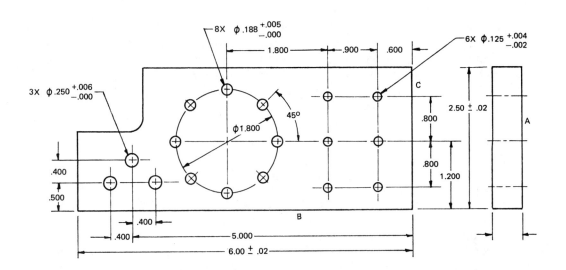

Unit Introduction

Today, programmed machine tools and CAM software, designs, and machined products have ever more complex shapes and functions. These advanced designs require closer, more technical inspection. This puts a particular challenge upon equipment and people performing the inspection. A major part of that challenge, today and for future improvement, lies in geometric theory.

Can we test these modern shapes with the accuracy with which they were manufactured? With widespread use of CAD/CAM and advanced CNC equipment, the industry is always faced with the quandary of having more complex, closer toleranced manufacturing abilities than measuring abilities.

What About the Future
It is hard to imagine any manufacturing process with any degree of sophistication that does not require advanced measuring and geometric concepts. This unit will show you that many of the solutions to better future metrology lies with geometric concepts. The goal will always be 100 percent accuracy—whether this goal can be met is theoretical. Achieving absolute accuracy is much like the story of a man trying to reach a wall by moving halfway each time he moves.

Moving halfway toward a wall, then halfway again, then halfway again repeatedly—each step becomes much more difficult to take. In theory, he will never get to the wall.

Absolute accuracy in evaluating part geometry is much the same. As our knowledge of geometrics and related inspection equipment becomes more accurate and sophisticated, the next step toward absolute accuracy becomes more complex and more refined than the last.

Chapter 17 will broaden your understanding of the theory and problems behind measuring. It will show you that we have research yet to do to answer some basic questions such as "How straight is this part?".

Due to the power of computers, the role of ANSI is expanding and the new updates are going to be very complex, challenging, and highly mathematical. We are closer to absolute accuracy today but must strive to be even closer in the future. There is much room for improvement in the way we measure part geometry, dimensions, and tolerances.

Chapter 18 will discuss another aspect of absolute accuracy: surface roughness and how, at the micro level, it has an effect on all element analysis. Geometrics and surface roughness become a single interwoven subject as we become ultra precise. Predictably, there will emerge a CMM or some other advanced measuring device that will evaluate roughness simultaneously with element analysis. Given tight enough tolerances, you cannot say with certainty how far a hole is from a datum without looking at the micro-finish of the hole, and the datum too, as an integral part of the deviation of the control.

CHAPTER 17

Geometric Element Analysis

There are challenges in evaluating most geometric controls. For example, in order to determine exactly how far a hole is from a datum surface requires the inspector to know exactly how flat the datum surface is and how round the hole is. *The entire process begins with element analysis of form.*

There is more to be learned about straightness, flatness, roundness, and cylindricity than has been revealed thus far. At the present time, there is truly no 100% accurate way to measure these elements for the true degree of variance from perfect.

It is a relatively easy challenge to determine if an element is within the form tolerance but quite another matter to say exactly how far an element may be from perfect.

Let's explore the concept of straightness one more time but from a theoretical standpoint. We will look at exactly what we are trying to prove. We will also look at what is being done today with computer inspection equipment and where the factors lie that cause the uncertainty. What needs improvement for future accuracy?

Some of the problems, as you might expect, are hardware. But some of the more interesting factors are in the mathematical definitions and computer methods used to rout out the geometric truth. One other problem blocking future accuracy is political—all the research agencies and companies working on solutions need to share data.

Today, all involved are working on more refined methods for determining exact values for element analysis. Government labs, universities, and industry are all working on solutions. Perhaps we will never reach the ultimate goal of 100% accuracy and certainty, but we are better off today than ever before and tomorrow will bring new solutions and even more accurate results.

Straightness—One More Time

In this first example, we will follow how a CMM goes about determining the straightness of an element. We will see why this procedure is so vital for measuring and why today we are still not accurate enough. Complex challenges exist when evaluating every geometric control with the exception of runout. Inspecting runout is absolute. There is no ambiguity with runout as long as the central axis is accurately aligned.

Computer measuring follows this procedure:

1. Collect points along the element (optical or touching).
2. Mathematically connect these points into an element line model.
3. Analyze the element—compare it to a perfect counterpart.

Each step has some problems that require future refinements—we will see what they are.

This straightness example has ramifications throughout the entire subject of geometrics. When we say "how far is a hole from a datum edge of a part?" we need to have some pre-answered data. How round is the hole and where is the exact center of the hole? How straight is the datum edge and where does the true datum lie? All of these questions and many others can be understood using the problem of straightness. Each question requires some analysis of an element or a surface.

First, what exactly are we saying when we ask "how straight is this element"? The truth that we attempt to prove is easy to define. Straightness is the amount of deviation of a single element away from a perfect straight line. The theoretical test of straightness then is the thinnest pair of perfect straight, parallel lines that contain the element in question. The thinnest possible "floating sandwich" containing the element in question is the absolute value of the straightness.

This is easy to understand but hard to prove. Let's look at two approaches. Both work, but only one could be 100% exact.

Element Analysis

Figure 17-1 is a highly exaggerated element on top of a part. Its shape has been carefully drawn to illustrate an important point. Element analysis is more complex than thus far revealed.

First, let's test it for straightness using what you already know. Remember, you learned that straightness is a floating control and thus the theoretical test templates must conform to the part as it is found. We slide a perfect straight template against the part and look for the maximum gaps. The distance from the template to the farthest gap is the value of straightness, right? Maybe not!

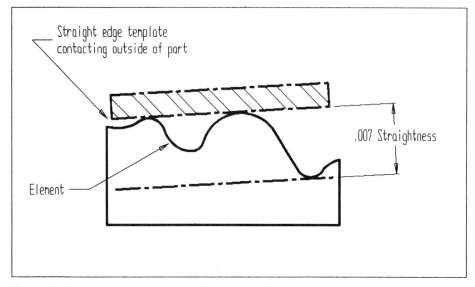

Figure 17-1 An exaggerated example of a straightness test using outside contact only.

By putting a perfect straight edge on the outside surface of the part, then looking for the largest deviation (point A), we have determined a possible false value. Suppose it is .007 inch from the top test line to the bottom. Although the straight edge test is common and useful, there is a subtle inaccuracy.

True Template Alignment

Are the outside bumps always the correct alignment points to determine the orientation of the thinnest possible theoretical sandwich? The answer is "not always."

Consider Figure 17-2. This is the same element but as you can see we have used the inner deviations to orient the theoretical sandwich. This produces a sandwich that lies at a slightly different angle to the part thus the sandwich is less than the .007 inch result in the previous test. By letting the inward bumps establish the orientation of the sandwich, we arrived at a different straightness value. Notice that both tolerance zones in Figures 17-1 and 17-2 contain the element, but only Figure 17-2 was the true *thinnest* sandwich that contained the element.

In other words, when we allow the sandwich to truly float—that is, align to the element in any fashion—it is possible to find thinner sandwiches than the one created when we locate a real straight edge on the *outside* of the part.

From this example you can see that the straight edge test was evaluating the

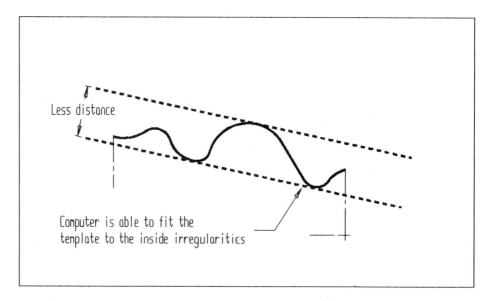

Figure 17-2 Using element analysis, we see that the thinnest sandwich may require template alignment other than the outside of the part.

element from the outside of the part and not analyzing the true element. Computers allow us the option of analyzing the element from a profile view as in Figure 17-2.

Element Analysis

To determine the true element sandwich requires two vital procedures:

1. *Collect Data* to get a clear picture of the actual shape of the element.
2. *Analyze the Element* for deviation away from perfect.

A modern CMM can perform both tasks. In fact, this is the major difference in what a CMM is doing when inspecting features of a part. It is geometrically analyzing elements before reporting control results. This is an elusive task that becomes difficult, if not impossible, using standard measuring methods.

From this example you can see that, before we can say for sure how straight an element actually is, any form element must be analyzed and a truly floating template sandwich fitted around it. Extending this element analysis example, you can see that flatness, roundness, and cylindricity, and the controls of profile as well, might have the same challenge when finding the true deviation from perfect. (Element analysis is simplified for profile if the control relates to a datum—then the answer is absolute with respect to some outside datum such as a layout table.)

In actual practice, this effect becomes less pronounced as the element comes close to perfect. But as we consider absolute accuracy, element analysis must be considered. Analysis is a central challenge in evaluating any other geometric dimension, control, and tolerance—for example, perpendicularity. The process must start with an analysis of the surface feature elements in question.

Computer Element Analysis

Now let's examine the CMM process for the methods and future improvements needed.

1. Collect points along the element.
2. Mathematically connect these points into an element line or surface model.
3. Analyze the element/model for control value.

To analyze the element requires a data base that represents the line. There are two ways to obtain the data; touching the part or not touching the part. These are referred to as "Tactile" and "Nontactile." Each has its own particular challenges; however, we will not discuss this aspect of accuracy. Here we are concerned with the geometric question.

Either method needs to produce enough accurate data that the element in question is truly visible as a floating entity—the true shape of the element in the control direction.

The first step is data sampling. Consider Figure 17-3. Here we are using tactile probe to gather the element data. Notice that the shape and size of the probe affects the data we collect. OK, fine, we get a smaller probe. Yes, that is an answer but it has limitations. At some point, as we get smaller probes and look more closely at the test point, we magnify our view until we run out of smaller probes. At this stage, surface roughness becomes a major factor in gathering element data. Chapter 18 discusses roughness and how it is being dealt with today.

So here are two challenges to absolute accuracy: 1) Sampling density, and 2) Surface roughness.

Step 2 is how does the CMM draw the element? Another problem is mathematical in nature. Suppose we have an accurate data base with enough points to truly represent the element in question. These are just points at this time—they are not a line. They must be connected within the computer to form the element line to be analyzed. There is yet another hurdle. The computer must use a process

238 □ Unit IV Geometric Theory

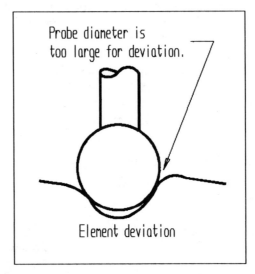

Figure 17-3 For tactile methods of data collection, the probe size can affect the process.

to actually derive the true element. It is a mathematical method and the tools used are called "Algorithms." Algorithms are formulas that define the way the points are connected. They are attempting to draw a true representation of the element in question using data points along the element. Unless the collected data is infinitely close together, it isn't possible to simply connect the points using straight lines. The line produced must represent the entire surface. They must be mathematically "splined."

Splining can be best understood through ship building. In ship building, the long planks on the side of the ship only take on the correct shape as they are fastened to the ribs. Each rib defines the overall curve of the board. A missing rib or one that protrudes too far will not only change the curve at the erroneous position, but will have a ripple effect on the adjoining parts of the curve. This represents a data sampling problem.

Splines are also used in drafting where they are long flexible plastic boards. The spline is held or fixed to a number of plotted points and the resulting curve is drawn. The computer simulates this splining concept. Using an algorithm, it attempts to draw an element line as though the spline was fixed at each point in the collected data.

Deriving an element from data points is close to accurate but not perfect! Depending highly on the density and accuracy of the data, and the algorithm method, the produced element can have bumps. Consider Figure 17 – 4. There are any number of possible ways of connecting the four points shown. Which is right?

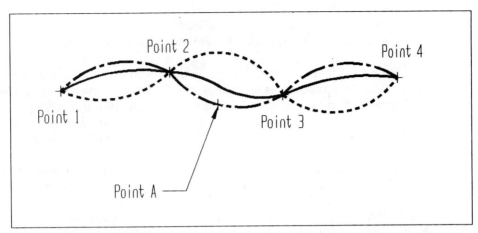

Figure 17-4 All three lines pass through points 1 – 4. Only one line passes through point A.

You could answer "choose data points that are closer together." For example, let's add point A. Now that helps to stabilize the curve. If we go through point A we get a better result. But this process has the same infinite smallness limit as the probe size—how dense must the data be? We can zoom in close enough to start the entire question over again; the points will look too far apart—infinitely small. When do we stop sampling the part surface for element data? When is the data dense enough to get results that produce a true element? There is much controversy over this question.

The algorithms each lab finds are derived as the result of research and are considered proprietary. At the time of this writing, no one single lab is thought to have the final answer!

All are close and working on it. Will they share data? This may be the only answer. Let's now examine the three-step process of analyzing an element for improvements needed:

1. Collect Data
   The probe size is a problem for tactile methods.
   The data density sampling is in question. How dense?
   The accuracy of the collecting hardware is also important
   Heat, Constant Repeatability, Stability
   Surface finish effect on data at a micro-level
2. Create the Element Model
   The mathematical methods—the algorithms—are questionable.
   Data density is also in question at this stage.

3. Analyze the Model Element
   This phase is somewhat solidified. Today we are reasonably sure that given an accurate element, we are able to analyze it with some certainty.

Am I saying that CMM work is not to be trusted today? No. The accuracy of industrial CMM equipment is amazingly reliable but we can see there is room for improvement. Accuracy is far better today than in the past but solving the problems listed above, we can see even more improvement for the future. These problems become technically challenging and more difficult as we approach absolute accuracy.

Again, using our straightness example, most of the problems associated with data collection and element generation become nearly nonexistent as the element becomes closer and closer to perfect. A truly straight element would not have a data density problem, nor a probe shift, except that resulting from the micro-finish. All involved agencies—ANSI, universities, government labs, and private industry—are working on improvements to the accuracy of metrology, but at the heart of the issue are the concepts of geometrics.

CHAPTER 18

Surface Roughness Evaluation

In the past, the study of surface roughness control was found only when product smoothness, appearance, or function of the surface such as the sliding action was of concern. There is another aspect to roughness—geometric form. In the effort to reach absolute accuracy, a major blockage is surface roughness. As control tolerances dip beneath .0005 inch, surface roughness must be considered as part of the geometric form of the feature in question. From Chapter 17 we learned that evaluation of any control starts by testing the form of both the feature in question and any related datums.

A machined finish of 125 micro-inches is considered a general machine finish but, as you will see in this chapter, the height of the micro-ridges to the bottom of the valleys will be over .0002 inch simply due to the finish. Finish alone can consume much of the control tolerance and affect the element analysis.

Historically, roughness and form have been considered as separate subjects when in fact they are not. Roughness is a degree of form. The surface roughness affects probe-collected data. A small probe touching different sides of a single surface irregularity will produce different form data depending upon which side of the bump is touched.

For both tactile and nontactile collection methods, the roughness is part of the surface and thus part of the features that create the form. While present roughness systems are mostly a surface smoothness control, in this chapter we are more concerned with the roughness as it affects element analysis and surface form.

The shape of each roughness peak on a part is of prime concern. This shape is a combination of the roughness height, the roughness width, and waviness. We will examine each soon. This material is not a lesson about surface finish, but rather it is designed to create an awareness of the effect finish can have on geometric element analysis. If you wish a more complete treatment of surface finish, it is suggested that you read *Machinery's Handbook* by Industrial Press.

The roughness becomes part of the data collection and definitely affects the final result. Classically, roughness is determined by moving a diamond stylus over the surface in question. Form is determined by moving a probe over the surface. Predictably, both data collection processes will be combined. This is a major barrier to 100% accuracy at this time.

Absolute accuracy requires a measuring process that simultaneously accounts for micro-irregularities in the elements as they are being analyzed. This is an area where much improvement is yet to be made on commercial inspection equipment.

You can envision that a small CMM probe touching a part would give different results depending on which side of a surface ripple it touched. At the present time, roughness can be evaluated on sophisticated equipment such as Figure 18–1. This is a stand-alone surface evaluation testing machine. It has various electronic filters to allow the user to see only that part of the surface roughness in question. The waviness and or roughness may be shown alone. We discuss this next. Also, large graphs of the surface irregularities are printed out for the user.

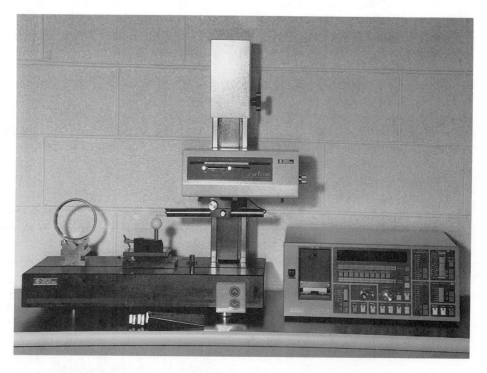

Figure 18–1 A microprocessor, stand-alone, surface finish machine. (Photo courtesy of Brown & Sharpe Corp.)

CHAPTER 18 Surface Roughness Evaluation   □ 243

However, as sophisticated as this may sound, surface finish data is only indirectly applicable as form data. The form data collection is a separate process on a different machine.

In this chapter we will examine four aspects of the micro-inch roughness system.

1. The meaning of micro-inch roughness figures—height and width
2. Comparison median lines
3. The averaging spans called the cutoff distance
4. Waviness and its relationship with roughness

Each of these may be found on the roughness symbol in Figure 18–2.

Surface Roughness Analysis

MicroInch System

The unit used to measure surface roughness height is the micro-inch—.000001 inch (one millionth of an inch). Roughness values are called out as average deviations from a median line in micro-inches. That is important enough to repeat. The roughness is specified or rather limited by an *average* deviation away from a median line. In Figure 18–3, you see a highly magnified surface element in question.

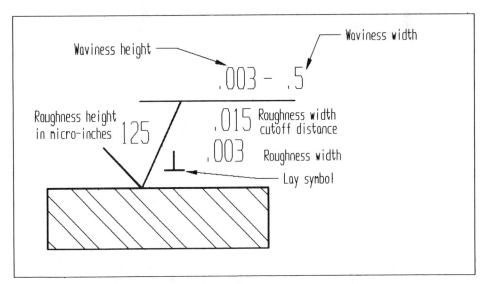

Figure 18–2 Several data fields exist on a micro-inch finish symbol.

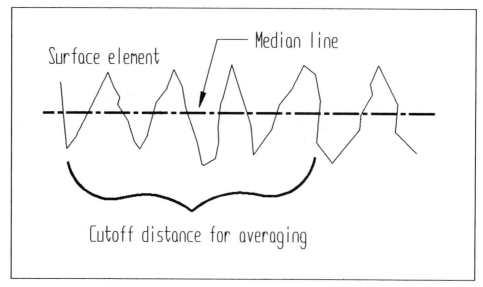

Figure 18-3 A roughness element.

Surface roughness is generally defined as "the smallest surface irregularities produced by or those irregularities affected by the manufacturing process." Even smaller surface irregularities are defined as material structure and cannot be changed by machining or abrading.

The smallest finish irregularities we can produce are in the one to two micro-inch range. This very fine finish is commonly achieved by abrasive action. The fineness of the finish is limited by the size of the abrasives available.

Roughness Average

Surface roughness testing is an averaging process. To determine the roughness, all the distances above and below a median line are recorded for a certain span. The median line is similar to a pitch diameter for a thread. It lies approximately halfway from the top of the peaks to the bottom of the valleys. The computed average deviation must not exceed the limit specified. The average of the high points and low points is taken over a distance called the cutoff range. We will discuss this median line more, later in this chapter.

In the process, all the peak heights and valley depths are averaged and the result is the surface roughness number in micro-inches. It is not the physical distance from the top of the peaks to the bottom of the valleys. The actual distance will be close to twice the roughness number.

A roughness specification of 125 micro-inches then is 125 micro-inches average over the cutoff range. Not .000125 inch from the top to the bottom of the surface irregularities.

Here we see why surface roughness can be of prime importance when evaluating feature elements. Suppose we have a position requirement of .0005 inch for a hole from a datum. Suppose the overall finish on the part is 125 micro-inches. That actually means a peak to valley depth of .00025 inch. If this is so for both the hole and datum feature, the roughness irregularities of both features can equal the entire position tolerance! This is worth thinking about!

Maximum Roughness is More Common

Occasionally, designers will be concerned with the minimum degree of smoothness for applications such as adhesive laminating or nonskid surfaces. This will then be expressed as a minimum value below the maximum roughness value shown on the crook of the bracket. However, far more common, you will find only a single value shown in the bracket. This then is the maximum roughness and any other lower (smoother) value is acceptable.

Roughness Width

This is the distance between irregularities. The maximum value acceptable is shown on the roughness symbol. Any value closer together is acceptable.

Cutoff Distance

The span over which the average deviation is taken is called the cutoff distance. It is chosen to contain a given number of irregularities. The smaller the cutoff, the more erratic an individual surface might appear in an examination. However, this is a machine function only; changing the cutoff only affects the averaging span, not the actual finish.

For example, If there was a single protrusion averaged over a large distance, it would have less effect on the result than when folded into a short cutoff distance. Common cutoff distances vary from .003 inch to .100 inch or .076 to 2.5 millimeters.

Waviness

A second surface phenomenon is that of waviness. Waviness is a cycle impressed upon the roughness. It is generally induced by vibrations or other cycles in the machining process. It can also be produced by warpage and heat-treating or other mechanical effects. You could envision waviness as a large ocean wave while roughness is the irregularities of the surface water upon the wave. It is undefined at this time, where waviness ends and form irregularities begin.

However, waviness and form are one and the same; they are both deviations of the surface being tested. Again, this calls for more effort being placed on integrated roughness-form testing equipment. Both roughness and waviness are deviations away from a perfect form.

The surface testing equipment in Figure 18-1 is designed so that various aspects of the surface may be individually highlighted. Waviness features may be filtered out of the roughness result to see details. Or the waviness could be viewed by itself by filtering out roughness.

Median Line

There are two methods used to determine the location of the median line in the roughness test. The first is a simple mathematical average of the heights of both valleys and peaks. This process is called "Arithmetical Averaging" and is often shortened to Arith or AA. The second median line placement method takes into account the total *area* above and below the line. It is called the Root Mean Square method or RMS. The RMS test tends to be more severe. In other words, a given surface will appear rougher using a RMS median average line than when tested using an arith line. When no callout for the averaging process is shown, the Arith method is commonly accepted.

The surface roughness symbol in Figure 18-2 is as it would appear on a print. This symbol is an evolution from a simple checkmark indicating that a certain surface was to be machined. The checkmark was expanded to include more information.

One aspect of surface finish, upon which we will only touch, is the "Lay." The lay is the direction, shape, and method by which the finish was achieved. For example, a ground part would have a different lay than end milling or peripheral milling. You can read more about the surface lay in *Machinery's Handbook*.

The actual required micro-inch finish is found in the crook of the bracket. The cutoff distance is above the cross bar. The waviness will be found inside the bracket along with the lay, should it be required. You should know that often only the roughness number is included. When this occurs, a cutoff of .030 inch is acceptable and there is no waviness control. This then indicates that the form control is a sufficient waviness control.

As we approach absolute accuracy, surface finish must be considered as part of a form deviation. Equipment is yet to be developed that considers both element and finish as part of the analysis.

A Final Word

The goals of this book have been to give you some real working skills in Geometric Dimensioning and Tolerancing; to give you the background to continue learning about geometrics from ANSI and from your own experience; and, above all, to give you a geometric consciousness about manufacturing. I can think of no other manufacturing subject that is more challenging nor more necessary to future developments in designing, machining, and measuring.

Your next lessons in geometric skills, accuracy, and understanding will take time. Very similar to improving measuring equipment, this first step will be easier than the next because further training for you will be more detailed. Time and experience will be your instructor. Thank you for being my student.

Glossary

Angularity—One of the orientation controls. To be within angularity control, either feature elements or centerlines/planes must be within a tolerance zone built around a perfect model that is at a specified basic angle with respect to a datum. The shape of geometric controls of orientation are the distance across a pair of parallel lines.

Balance Sheet—The comparison of the control tolerance earned to the control tolerance consumed. Referred to as *Burned* versus *Earned* in this book.

Basic—A perfect target dimension of the expected result. Basic dimensions have no tolerance. Working tolerance is then built around the basic dimension.

Bonus Tolerance—Additional tolerance that is earned by a feature deviation away from its worst case fit (MMC or LMC). Bonus is only possible after machining. The bonus equals the amount of deviation.

Circular Runout—Control of surface wobble of rotating objects, as compared to a datum axis. Runout is detected by indicating the rotating object. Runout also controls the form of the object.

Composite Control—A control that regulates more than one characteristic or built from more than a single control. See *Cylindricity*.

Concentricity—A control of location. Concentricity controls the axis of a particular feature as compared to a datum axis established by another feature.

Cylindricity—A composite control that regulates roundness, straightness, and angularity/parallelism of surface elements on round objects.

Datum—A geometric reference for measurement, dimensioning, and position. Datums are theoretically perfect lines or surfaces established by part features but they are not the feature. Datums represent the assembly world around the part feature.

Datum Feature—Any part feature that establishes a datum. The datum becomes a composite of the feature with its associated irregularities.

Datum Frame—A complete set of datums that define the part in three dimensional space. A datum frame on a print becomes a fixture and tooling points on a real part.

Datum Targets—Specific locations on the print where contact is to be made to hold the part for machining or inspection. Datum targets become tooling points in the shop.

Deviation—Any machined size that is different from the MMC/LMC size within the tolerance range.

Element—A surface line running in the control direction. You could draw an element on the surface of a part. For the part to be within geometric control, its elements must fall within the control zone. There are an infinite number of elements on any surface.

Feature—Any part of a design or product that can be dimensioned, toleranced, and measured. Holes, surfaces, and threads are all examples of features.

Feature of Size—Any part feature that has an associated size range. Often a feature of size is designated as a datum feature. Therefore the established datum has some size range too.

Feature Control Frame—The box drawn around a given control on a print. Much of the information on a geometric design is communicated within the feature control frames.

Floating Fastener—A method of finding the position tolerance for features such as drilled holes when they are to be assembled with fasteners such as rivets.

Fit—The real way machined products go together. One of the two basis for the geometric dimensioning and tolerancing system. See *Function*.

Fixed Fastener—A method of calculating position tolerance for features such as drilled holes where they will be assembled using fasteners that do not self align. Examples of fixed fasteners would be threaded bolts or pressed pins.

Form Controls—Straightness, roundness, flatness, and cylindricity are controls of form. These control the surface elements of features within a tolerance zone without datum reference. See *Floating Control*.

Floating Control—A control that conforms to the surface as it is found, without reference to a datum. Controls of form and sometimes profile are floating controls.

Function—the real way designed features work. One of the two methods by which geometric dimensioning and tolerancing is determined. See *Fit*.

Functional Gage—A tool that simulates the extremes of fit for a part feature. Functional gages save time and ensure correct fit with a minimum of measuring. Similar to a go-nogo gage.

Geometric Error—The distance across the consumed error zone created by mis-positioned feature centerlines. The geometric error is twice the radial error.

Geometric Tolerance—The allowable deviation away from the perfect model. This will always be the distance across a zone built around the perfect model of the desired thirteen characteristics.

Least Material Condition—One of two possible fit conditions. LMC represents the loosest fit within any design tolerance. LMC designs are most often concerned with controlling thinout of some wall. LMC became official in the 1982 ANSI standards.

Location Controls—These include position and concentricity. Both control the center of features with respect to a datum. Bonus tolerance may apply to position but not to concentricity.

Maximum Material Condition—The tightest fit possible within any design tolerance. Bonus tolerance may apply to any machined feature deviation away from MMC size.

Orientation Controls—A control of the angular relationship of either a feature's elements or its derived center with respect to a datum plane or line.

Parallelism—One of the orientation controls. To be parallel, either feature elements or centerlines/planes must be within a tolerance zone built around a perfect model that is at an angle of 0 (zero) degrees with respect to a datum.

Perpendicularity—One of the orientation controls. To be within perpendicularity control, either feature elements or centerlines/planes must be within a tolerance zone built around a perfect model that is at an angle of 90 degrees with respect to a datum.

Position Tolerance—The control of a feature's center with respect to a datum.

Profile Control—A surface control for objects that have a simple (single curvature limit) shape. Very similar to form except that the profile often is defined from a datum.

Regardless of Feature Size—RFS means that functionally, there can be no bonus tolerance. Whatever the deviation in feature size, the tolerance remains as stated on the print.

Rule of Application #1—PERFECT FORM ENVELOPE—Unless otherwise permitted by a note, when no form tolerance is specified for a feature that is to machined to size, the form will be limited at the perfect form envelope.

Rule of Application #2, Pre 1982—MMC bonus tolerance must be specified for all controls other than position. RFS is automatic unless otherwise specified. MMC bonus is automatic for controls of position.

Rule of Application #2, After 1982—MMC or LMC bonus must be specified for all controls other than position. RFS is automatic unless otherwise specified. For controls of position only, either LMC, MMC, or RFS must be shown in the feature control frame.

RC Distance—The amount of *radial change* needed to rework a feature center back within position tolerance.

RE Radial Error Distance—The actual physical difference in a feature centerline and the perfect position. The RE is one half a features geometric error.

Runout—Control of surface wobble of rotating objects as compared to a datum axis. Runout is detected by indicating the rotating object. Runout also controls the form of the object.

Straightness Control—A form control of surface elements within a tolerance zone. Special case—straightness—may apply to a derived center of a feature.

Tooling Points—Specific locations on the part where contact is to be made to establish datum reference.

Total Runout—A composite control of a surface. Total runout is an all element control (3 dimensional) of form with respect to a Datum.

Tolerance—The allowable deviation away from perfect.

Virtual Condition—The total resultant envelope any feature may consume within its design. To test any design, the virtual condition must be analyzed for assembly; it is the absolute composite, worst case for fit.

Appendix

- Challenge Problem
- Rework Sheets and Formulas

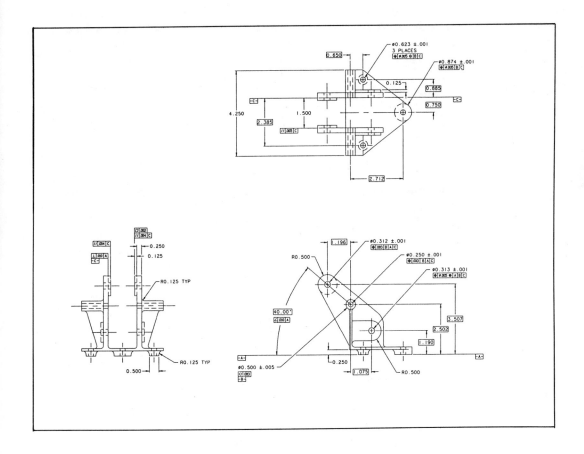

GEOMETRIC POSITIONAL REWORK

Correction Facts

What is the Balance amount needed? _____

What is the feature size change? _____

Radius change is 1/2 of feature size.   RC=_____

Feature C/L   Feature C/L
X Error = _____   Y Error = _____   Geo Radius = _____ GR

Correction will be: [ ] Polar   [ ] Rectangular

Polar Method

GA=   RC=

TAN GA=Y/X

SIN GA=Y/GR

COS GA=X/GR

Rectangular Method

XC=   YC=

$$XC = \frac{X \times RC}{GR}$$

$$YC = \frac{Y \times RC}{GR}$$

Rework Sketch

Sketch Error and Corrections in Correct Quadrant

Figure 14–10 Rework worksheet.

GEOMETRIC POSITIONAL REWORK

Correction Facts

What is the Balance amount needed? _____

What is the feature size change? _____

Radius change is 1/2 of feature size.   RC=_____

Feature C/L   Feature C/L
X Error = _____   Y Error = _____   Geo Radius = _____ GR

Correction will be: [ ] Polar   [ ] Rectangular

Polar Method

GA=   RC=

TAN GA=Y/X

SIN GA=Y/GR

COS GA=X/GR

Rectangular Method

XC=   YC=

$$XC = \frac{X \times RC}{GR}$$

$$YC = \frac{Y \times RC}{GR}$$

Rework Sketch

Sketch Error and Corrections in Correct Quadrant

Figure 14–10 Rework worksheet.

Index

Absolute value, 92
Algorithms, 156, 238
ALL element controls, 75
American National Standards Institute (ANSI), 8
   rules for controls of form, 47-49
Angularity
   angle tolerances not fan-shaped, 84-85
   control angle defined by basic dimensions, 83
   definition of, 82-83
   inspecting controls of, 84
   testing with template and optical methods, 85-86
Arithmetical Averaging, 246

Balance sheet, 175-81
Basic dimensions, 83, 144
Bidirectional tolerance zone, 134
Bonus tolerance, 29, 88, 92, 121
   from datums of size, 216-18
   double
      from group of features, 220-24
      from single feature, 218-19
   functional fit as basis for, 89-90
   none if RFS symbol present, 88-89
   reworking features where applicable, 163-66
Burned tolerance, 154, 157

Center features, 41
Centerline, 41
datum, 19-20
   deriving from feature, 122-24, 155
   differential movement of, 225
   of feature of size and control, 91-92
   orientation control for, 75
   position control for, 117
Centerplanes, 41
   position control of, 117
Circular runout, 103, 105
Coaxiality, 109-10
Complex tolerances
   bonuses from datums of size, 216-18
   double bonus
      of group of features, 220-24
      of single feature, 218-19
   in rework, 224-29
Composite positional tolerance, 133
Computer measuring machine (CMM), 61, 65, 77-78, 81-82, 113
   algorithms for feature irregularities, 156
   straightness of an element and, 234-40
Computer profile testing, 114
Concentricity
   function of, 110-11
   inspecting, 111-15
   review of, 116
   for verifying coaxiality, 109-10
Conical tolerance zone, 134
Conversion charts, 175
Coordinate dimensioning, 118-20
Coordinate measuring machine (CMM)

258 □ Index

for flatness, 55
for straightness, 45, 47
Correction methods
  polar, 196-99
  rectangular, 190-96
  RFS compared with MMC, 184
Cutoff distance, 244, 245
Cylindricity
  inspecting, 63
  review of, 63-64
  tolerance zone for, 62

Datums
  callout symbols, 19-20
  compound, 22-23
  definition of, 23, 31
  features in establishing, 17-19
  frames, 26-30
  implied, 20, 30
  importance of, 15
  primary, 20
  priorities for controls, 20-23
  reference within pattern, 132-33
  review of, 31-32
  of size, 29, 216-18
  target symbols, 23-26
  temporary, 25-26
  understanding of, 15-17
Design requirements
  fixed fastener for positional tolerances, 210-15
  floating fastener for position tolerance, 207-10
  geometric priorities linked to function, 202-204
  virtual condition and, 204-207
Drawings, reading, 36-37

Earned tolerance, 90, 92, 157
  calculating, 96-97, 121-22
Elements, 40-41
  analysis of form, 233-40

in profile controls, 66-67
Embodied controls, 62
Error
  correction methods RFS and MMC, 184
  direction of, 173
  Geometric Positional Error, 150, 158, 175-81
  line of, 174
  polar correction method, 196-99
  Radial Error Distance, 172-73
  rectangular correction method, 190-96

Feature
  control types, 6-7
  datum, 17-19
  definition of, 17
  irregularities in, 156
  out-of-balance, 156-57
  single axis, 149-71
  two axis, 172-99
Feature control frames
  combined form, 34
  information found in, 35
  multiple form, 34
  scanning a print, 36-37
  single form, 33-34
Feature Size Change (FSC), 182-83
FIM test, 77
Fixed fastener assembly, 210-15
Fixed template, 66
Flatness
  determining, 54
  embedded control of, 76-77
  flatness trap, 52, 54
  inspecting, 52
    CCM, 55
    indicator/inspection table test, 54
    inverted method, 55
    leveling method, 54-55
    optical flats, 55-56
    optical laser, 56
  review for, 56

tolerance zone for, 51-53
Floating fastener assembly, 207-10
Floating template
   for controls of form, 43-45, 54, 57, 63
   for controls of profile, 65-66
Floating zone, 75-76
Form controls, 6
   ANSI Rules of Application, 47-49
   CMM test for straightness, 45
   cylindricity, 62-64
   flatness, 51-56
   floating template control, 43-45
   roundness, 57-62
   straightness, 45-51
      of centerlines or planes and, 42-43
   2D/3D versions, 43
   See also Profile controls
Free state variation inspection, 136
Functional fit, 16, 48-49
   bonus tolerance and, 89-90
   geometric priorities and, 202-4
Functional gage, 213

Geometric control, 41
Geometric dimensioning, 118
Geometric Dimensioning and Tolerancing system, 139-40
Geometric Error (GE), 150, 158, 175-81
Geometric hole, 78
Geometric position, 90
Geometrics
   advantages of, 8-12
   definition of, 3-4
   feature control of, 6-7
   statistical process control and, 5-6
   symbols in, 10, **11**

Hole
   boring along error line, 183-84
   computing balance sheet for, 175-81
   position control for, 118
   reaming at MMC, 183

Indicator layout table test, 43-45, 46, 54
Inspection
   concentricity, 111-15
   cylindricity, 63
   flatness, 52, 54-56
   geometric position, 122-24
   parallelism, 77-78
   perpendicularity, 80-82
   position of single axis feature, 149-56
   roundness, 57, 59-61
   runout, 105-6
   straightness, 43-47, 234-40
   two axis features, 172-81
Interferometry, 56
Inverted method, 55

Lay, 246
Least Material Condition (LMC), 93
   position using, 120-21
   reworking features at, 164
   symbol for, 88
Leveling method, 54-55
Line profile
   inside mold line (IML) application, 68
   inspection of, 69-71
   normal line, 70-71
   outside mold line (OML) application, 68
   tolerance zone for, 67-68
Location control, 6, 109
   concentricity, 109-16
   position control, 116-28

Machinery's Handbook, 241, 246
Material conditions, 92, 117
   Least Material Condition (LMC), 93
   Maximum Material Condition (MMC), 93
   Regardless of Feature Size (RFS), 100-2
   review of, 102
   symbols for, 88
Mating parts, 210-15

Maximum Material Condition (MMC), 93
  perfect form envelope at, 48-49
  position using, 120-21
  rework of feature at, 157, 164, 166
  symbol for, 88
  zero position tolerance at, 130-32
Median line, 246
MicroInch System, 243-44
Micrometer test, 59

Natural function, 4-5
No bonus tolerance, 88-89
  reworking features where applicable, 157
No form specified (ANSI rules), 48
Nongeometric tolerancing. See Coordinate dimensioning
Nontactile data collection, 241
Normal line, 70-71

Optical comparitor, 85-86, 114
Optical flats, 55-56
Optical laser, 56
Orientation controls, 6
  angle of control to the datum, 74
  angularity, 82-86
  center feature symbols, 75
  parallelism, 75-78
  perpendicularity, 78-82
  review of, 86-87
  single or ALL element controls, 75
  tolerance zone shapes, 75

Parallelism
  definition of, 75
  embedded control of flatness, 76-77
  floating zone for, 75-76
  inspection of, 77-78
Part features
  coaxiality of verified by concentricity, 109-10
  datums from, 19

Perfect form envelope (ANSI rules)
  at MMC, 48-49
  at MMC not required, 49
  computer assisted evaluation, 49
Perfect position model, 172
Perpendicularity
  definition of, 78
  inspection of, 80-82
  tolerance zone of, 80
Photo contact master (PCM), 67
Polar correction method, 196-99
Position control, 109
  basic hole position, 118
  bonus tolerance only when MMC or LMC symbol present, 121
  for centerlines or centerplanes only, 117
  converting from rectangular to geometric position, 172-74
  coordinate tolerance zones smaller than geometric zones, 118-20
  datum reference always, 117
  definition of, 116
  earned tolerance formula, 121-22
  fixed fastener method for calculating, 210-15
  floating fastener method for calculating, 207-10
  geometric target zone, 118
  inspecting geometric position, 122-24
  material condition modifiers may apply, 117
  review of, 128
  RFS applied, 182-83
  tolerance zone shape for, 116
  2D aspect of, 118
  using MMC or LMC, 120-21
Prints, reading, 36-37
Priorities
  in datums, 20-23
  function and, 202-4
Profile controls, 42
  distinguished from form controls, 64-65

elements in, 66-67
fixed template for shape from datum reference, 66
floating template for shape without datum reference, 65-66
of line, 67-71
review of, 72-73
shape defining methods, 67
of surface, 71-72
tolerance zone of, 64
2D/3D versions, 43
See also Form controls
Projected tolerance zone, 129-30
    inspection of controls, 130
Pythagorean Theorem, 176-77

Radial Error distance (RE), 149, 172-73
Reaming, 174, 183
Rectangular correction, 190-96
Rectangular tolerance zone, 118-20
Regardless of Feature Size (RFS)
    after 1982 standards, 101
    automatic for all controls other than position, 100
    no bonus tolerance, 100
    position control applied to, 182-83
    pre-1982 standards, 101-2
    symbol for, 88-89
Restrained state inspection, 136
Rework
    complex, 224-29
    determining amount of, 181-87
    determining direction and amount of, 187-90
    LMC Method for MMC feature, 169
    LMC test formula for, 170
    review of, 200-1
    of single axis feature, 156-61
    of two axis feature, 172-99
Root Mean Square (RMS) test, 246
Roughness. See Surface roughness
Roundness

inspecting for, 57
    CMM, 61
    geometric sense required, 61
    micrometer test, 59
    rotating the part between centers with indicator, 60-61
    tri-anvil micrometer test, 59
    vee-block and indicator, 59
review of, 61-62
tolerance zone for, 57
Rule of Sides, 157-60, 182, 188, 189
Runout, 6
    circular, 103, 105
        inspecting, 105
        review of, 107-8
        RFS always applied in, 106-7
    total circular, 105
        inspecting, 106
        pre-1982 symbol, 105
        surface types controlled by, 106

Shapes
    defining, 67
    simple, 64, 72
Sine plate, 84
Single axis feature
    inspection of position, 149-56
    review of, 170-71
    reworking, 156-60
Splining, 238
Stack height, calculating, 84
Stated tolerance, 92
Statistical process control, 5-6
Straight edge template test, 44, 47
Straightness, 45
    ANSI Rules of Application, 47-49, 51
    bi-directional, 51
    of centerlines or planes, 42-43
    inspecting, 46-47
        CMM test for, 45, 234-40
        indicator layout table test for, 43-45
        straight edge test for, 44

review of, 49, 51
tolerance zone for, 43
Surface controls, 40-41
Surface profile
    simple shapes only, 72
    tolerance zone of, 71-72
Surface roughness
    cutoff distance, 245
    data collection and, 241-43
    definition of, 244
    form and, 241
    maximum roughness is common, 245
    median line location, determining, 246
    MicroInch System, 243-44
    roughness average, 244-45
    roughness width, 245
    waviness on, 245-46
Symbols
    datum callout, 19-20
    datum target, 23-26
    feature control, 33
    geometric characteristic controls, **138**
    for geometrics, 10, **11**
    for material condition, 88
    for symmetry, 8-9

Tab, inspecting, 149-51
Tactile data collection, 241
Tolerances
    bonus, 29, 88-90, 92, 121, 163-66
    burned, 154, 157
    cost vs., 10-12
    earned, 90, 92, 96-97, 121-22, 154, 157
    fixed fastener method for calculating, 210-15
    floating fastener method for calculating, 207-10
    stated, 92
    See also Complex tolerances

Tolerance zones
    composite positional, 133
    conical and bidirectional, 134-35
    for controls of form and profile, 42, 64
    for controls of orientation, 75
    entities found in, 39
    link to feature size, 90-91
    projected, 129-30
    2D/3D surface controls, 40-41
    zero position at MMC, 130-32
Tooling points, 23
Total circular runout, 105-6
Tri-anvil micrometer, 59
Tricoidal shape, 59
True template alignment, 235-36
Two axis feature
    computing a balance sheet, 175-81
    converting from rectangular to geometric position, 172-74
    polar correction method, 196-99
    rectangular correction method, 190-96
    rework of
        amount of, determining, 181-87
        direction and amount of, determining, 187-90
        review of, 200-1
2D position control, 118
2D/3D
    controls of form and profile, 43
    surface controls, 40-41

Vee-block and indicator, 59
Vernier bevel protractor, 85
Virtual condition, 18, 204-7

Waviness, 245-46

Zero position tolerance
    at MMC, 130-32